Springer Tracts in Natural Philosophy

Volume 18

W0037182

Edited by B. D. Coleman

Co-Editors: R. Aris · L. Collatz · J. L. Ericksen

P. Germain · M. E. Gurtin · M. M. Schiffer

E. Sternberg · C. Truesdell

Jürg T. Marti

Introduction to the
Theory of Bases

Springer-Verlag Berlin Heidelberg GmbH 1969

Jürg T. Marti

Department of Mathematics
University of Illinois, Urbana

ISBN 978-3-642-87142-9   ISBN 978-3-642-87140-5 (eBook)
DOI 10.1007/978-3-642-87140-5

To Rita

Preface

Since the publication of Banach's treatise on the theory of linear operators, the literature on the theory of bases in topological vector spaces has grown enormously. Much of this literature has for its origin a question raised in Banach's book, the question whether every separable Banach space possesses a basis or not. The notion of a basis employed here is a generalization of that of a Hamel basis for a finite dimensional vector space. For a vector space X of infinite dimension, the concept of a basis is closely related to the convergence of the series which uniquely correspond to each point of X. Thus there are different types of bases for X, according to the topology imposed on X and the chosen type of convergence for the series.

Although almost four decades have elapsed since Banach's query, the conjectured existence of a basis for every separable Banach space is not yet proved. On the other hand, no counter examples have been found to show the existence of a special Banach space having no basis. However, as a result of the apparent overconfidence of a group of mathematicians, who it is assumed tried to solve the problem, we have many elegant works which show the tight connection between the theory of bases and structure of linear spaces. In the more general setting of a separable locally convex topological vector space or a complete linear metric space, the basis problem is now solved; there actually are examples of such spaces which have no basis. By the nature of the problem, the methods of proof used in the theory of bases are those of functional analysis.

A few conditions sufficient for a sequence to form a basis of a certain type are now known. Moreover bases have been constructed for most of the separable Banach spaces which are presented as examples in text-books on topological vector spaces. For instance, the trigonometrical system is a basis for the Hilbert space $L_2[0,2\pi]$. On the other hand, if one assumes the existence of a basis for an abstract Banach space X, one obtains valuable indications on the structure of X or of closed linear subspaces of X. The assertions may concern weak sequential completeness, separability, reflexivity, dimension, or weak conditional compactness of bounded sets.

Some generalizations of the concept of a basis for a Banach space have greatly enriched the theory. On the one hand, it is natural to see how results on Banach spaces may be generalized to complete linear metric spaces or locally convex topological vector spaces. The definition of a basis, on the other hand, may be generalized itself. If one replaces the elements of a basis by linear subspaces of a Banach space or an F-space, one obtains decompositions of these spaces, and these decompositions exhibit some properties similar to those of the bases. It is worth noting that decompositions exist for every Banach space of infinite dimension. Carrying the generalization of a basis one step further, one first discards the idea of a series expansion, requiring of a biorthogonal system $\{x_\lambda, f_\lambda\}$ only that $\{x_\lambda\}$ is total in the space X; such a system is called a dual generalized basis. Moreover, if $f_\lambda(x) = 0$ for all λ implies that $x = 0$ for all x in X, the dual generalized basis is said to be a Markushevich basis. Finally, when the requirement of totalness, common to all definitions of bases and decompositions described above, is dropped along with the requirement of countability, we obtain the concept of a generalized basis for a topological vector space X.

Although there are presently more than two hundred publications on the theory of bases, up to now no text-book has been issued which collects and systematizes the essential results of the subject. This tract is an attempt to meet such a need.

The first chapter contains a short introduction to the working tools from functional analysis. The theorems are given there without proofs. This is justified, since there are now many well-written standard works on this subject. In Chapter II the fundamental theorems on unconditional and absolute convergence of series in Banach spaces are derived. Definitions and properties of the most important types of bases for Banach spaces, together with examples of bases in some well-known spaces of this type are given in Chapter III. The fourth chapter deals with the connections of bases, projections, orthogonality and simple $\mathcal{N}_1$-spaces, as well as with equivalent bases for Banach spaces. The known facts on the structure of Banach spaces with bases are explained in Chapter IV, and the following chapter is concerned with bases for Hilbert spaces. Decompositions are introduced in Chapter VII, and the application of both bases and decompositions in the theory of B-algebras, including compact operators, operators of finite rank, and the theory of proper π-rings, are discussed in Chapter VIII. Some of the most interesting results on generalized bases for topological vector spaces are presented in the final chapter.

I thank the Editor, Professor B. D. Coleman for his friendly co-operation and the Springer-Verlag for the careful preparation of this book.

Urbana, Illinois, February 1969   JÜRG T. MARTI

Contents

CHAPTER I

Linear Transformations

In the four paragraphs of this chapter we present some basic definitions and facts from functional analysis, as well as applications in special spaces. These preliminaries will be used in the subsequent chapters. Since many introductions to functional analysis are now available, in order to save space, we omit proofs of all of the lemmas, theorems and corollaries given here. Moreover, one will find here only the working tools which are really needed for the development of the theory of bases. We begin by defining various abstract spaces, and we list their most important properties. Then we investigate linear transformations of one space into another, continue with some facts on conjugate spaces, and conclude with results for several spacial spaces.

It is supposed that the reader is familiar with the notions of a field, a linear space (also called a vector space or a linear vector space), a subspace of a linear space, an algebra, and with some elementary definitions from the theory of sets and the theory of measure and integration.

1. Linear Topological Spaces

A *mapping* (sometimes also called *map, function, transformation* or *operator*) $f: X \to Y$ from (of) a set X into a set Y is a rule which assigns to each member x of X a unique member $f(x)$ in Y. For any subset A of X we write $f(A)$ to denote the set $\{f(x)|x \in A\}$. X is called the *domain* of f and $f(X)$, the image of X under f, is called the *range* of f. If $f(X) = Y$ we say that f is a function of X *onto* Y. f is *one-to-one* if and only if $f(x_1) = f(x_2), x_1, x_2 \in X$ always implies $x_1 = x_2$. For any subset A of Y let us denote $f^{-1}(A) = \{x \in X | f(x) \in A\}$. $f^{-1}(A)$ is called the *inverse image* of A. In particular, the inverse image of a single point y in Y will be denoted by $f^{-1}(y)$. If f is one-to-one and onto, $x = f^{-1}(y)$ defines a function f^{-1} of Y onto X, called the *inverse* (function) of f. If A is any subset of X, the real function χ_A, defined by $\chi_A(x) = 1, x \in A$, and by $\chi_A(x) = 0, x \notin A$ is called the *characteristic function* of A.

A *topology* for a set X is a family τ of subsets of X, called *open sets*, such that the void set, the set X, the union of arbitrary many open sets, and the intersection of finitely many open sets are open (i.e. are in τ). The set X, endowed with the topology τ is a *topological space*. If τ and τ' are two topologies for X and if $\tau \subset \tau'$, then τ is said to be *weaker* than τ', and τ' is said to be *stronger* than τ. τ is *equal* to τ' if and only if $\tau = \tau'$. A subset A of X is *closed* if its complement with respect to X is open. The intersection $\bar{A}$ of all closed sets containing a subset A of X is called the *closure* of A. A subset B of A is *dense* in A if $A \subset \bar{B}$. X is *separable* if there exists a countable dense set in X. A topological space Y is a (topological) *subspace* of X if and only if $Y \subset X$ and the sets which are open in Y are precisely the intersections of Y with the open subsets of X. A *neighborhood* of a set A in X is a set containing an open set which contains A. A collection B of open sets of X is a *base* (for the topology of X) *at the point* x in X if for any neighborhood N of x, there exists an open set A in B such that $x \in A \subset N$. A mapping f of X into another topological space Y is *continuous* if and only if $f^{-1}(A)$ is open in X for every open set A in Y. A one-to-one mapping f of X onto Y such that $f(A)$ is open in Y if and only if A is open in X, is called a *homeomorphism*. If such a homeomorphism exists, X and Y are said to be *homeomorphic*.

If Λ is a set in which a relation $\leq$ is defined with the following properties: (i) if $\lambda \leq \mu$ and $\mu \leq \nu$, then $\lambda \leq \nu$; (ii) $\lambda \leq \lambda$ and (iii) for $\lambda, \mu \in \Lambda$ there is a ν in Λ such that $\lambda \leq \nu$ and $\mu \leq \nu$; then Λ is said to be a *directed set*. The notation $\lambda \geq \mu$ is equivalent to $\mu \leq \lambda$. A *net* $\{x_\lambda\}$ in a topological space X is a map of a directed set Λ into X. If Λ is the directed set of integers $i = 1, 2, \ldots$, then, evidently, $\{x_i\}$ is a *sequence* in X. A net $\{x_\lambda\}$ in X *converges* to a point x in X written $x_\lambda \to x$ or $x = \lim_\lambda x_\lambda$ if to every neighborhood N of x there corresponds a μ in Λ such that $x_\lambda \in N$ for every $\lambda \geq \mu$. If such a point x exists it is called a *limit*. A topological space X is a *Hausdorff space* if and only if for distinct points x and y in X there are disjoint neighborhoods of x and y in X.

Lemma 1. *A toplological space is a Hausdorff space if and only if every net has at most one limit.*

A net $\{x_\lambda\}$ in a topological space X has a *cluster point* x in X if for every neighborhood N of x and $\mu \in \Lambda$ there is a $\lambda \geq \mu$ for which $x_\lambda \in N$. A closed subset A of X contains the cluster points of all nets in A. A subset A of X is *compact* if and only if each net in A has a cluster point in A. A is *sequentially compact* if and only if every sequence in A has a cluster point; in other words if and only if each sequence in A has a subsequence which converges to a point of X. A is *conditionally compact* if and only if $\bar{A}$ is compact.

Let Φ always denote the field of real (or complex) numbers $\mathbb{R}$ (or $\mathbb{C}$ respectively) and let X be a linear space over Φ. Elements of Φ are called *scalars*. A finite sequence $x_1, \ldots, x_n$ in X is *linearly independent* if and only if $\sum_{i \leq n} \alpha_i x_i = 0$, with α_i in Φ, implies $\alpha_1 = \alpha_2 = \ldots \alpha_n = 0$. A subset B of X is a *Hamel basis* for X if and only if every x in X has the unique representation $x = \sum_{i \leq n} \alpha_i x_i$ with α_i in Φ and x_i in B, where n is an arbitrary but finite integer which may depend on x. A finite number of elements in B is necessarily linearly independent. A Hamel basis for X always exists and all Hamel bases have the same cardinal number. This cardinal number defines the *dimension* of X. If there is a finite Hamel basis for X, X is said to be *finite dimensional* and the Hamel basis is simply called a *basis* for X. The set of all finite linear combinations of points in a subset A of X is denoted by sp A, the *span* of A, and the closure of this linear subspace of X is denoted by $\overline{\text{sp}}\, A$. If Y is a linear subspace of X, the *factor space* X/Y is the set of all sets of the form $x + Y (= \{x + y | y \in Y\})$ with x in X. The algebraic operations in X/Y are defined by the equations

$$(x + Y) + (y + Y) = (x + y) + Y, \qquad x, y \in X,$$

$$\alpha(x + Y) = \alpha x + Y, \qquad \alpha \in \Phi, x \in X.$$

With the operations thus defined, the class X/Y is a linear space. Moreover, an *isomorphism* of X into another linear space Y is a one-to-one map f of X into Y such that $f(\alpha_1 x + \alpha_2 x_2) = \alpha_1 f(x_1) + \alpha_2 f(x_2)$, $\alpha_1, \alpha_2 \in \Phi, x_1, x_2 \in X$. X and Y are said to be *isomorphic* if and only if there exists an isomorphism of X onto Y.

A *linear topological space* is a linear space X with a topology such that addition and scalar multiplication are continuous simultaneously in both variables. Consequently, for each x in X, each open (closed) set A in X and each nonzero α in Φ, the sets $x + A$ and αA are open (closed). A *local base B* in X is a base at the point 0; and it is clear that then $x + B$ is a base at the point x for each x in X. Two linear topological spaces X and Y over the same field are called *topologically isomorphic* if there exists a linear map of X onto Y which is a homeomorphism; if there exists such a map (which is a continuous isomorphism of X onto Y with continuous inverse), it is called a *topological isomorphism*. A set Y is a *linear* (topological) *subspace* of X if and only if Y is both a linear subspace of the linear space X and a topological subspace of X.

Lemma 2. *If X is Hausdorff, then every finite dimensional linear subspace of X is closed.*

If A is a subset of X, then it is known that $\overline{\text{sp}}\ A$ is a (closed) linear subspace of X. A is *total* in X if and only if $\overline{\text{sp}}\ A = X$. A is *bounded* if and only if for each neighborhood N of 0 there is a positive real number t such that $A \subset tN$.

Lemma 3. *A compact subset of a linear topological space X, and hence also a convergent sequence, is bounded.*

A subset A of X is said to be *symmetric* if $-x$ is in A whenever x is. A is said to be *convex* if $x, y \in A$ always implies $tx + (1-t)y \in A$ for every real number t in the interval $[0,1]$. A is said to be *circled* if $\alpha A \subset A$ for every α in Φ with $|\alpha| \leqslant 1$ and A is said to be *absorbing* if for any x in X there exists an $\varepsilon > 0$ such that εx is in A. If the collection of convex neighborhoods of 0 is a base at the point 0 (i.e. a local base in X), then the topology of X is said to be *locally convex* and X is called a *locally convex space*. A locally convex (linear topological) space is called *barrelled* or a *barrel space* if each barrel is a neighborhood of 0, where a *barrel* is a closed convex circled absorbing set in X.

Lemma 4. *Each neighborhood of 0 in a linear topogical space X contains a circled absorbing neighborhood of 0 in X.*

A *metric* on a set X is a non-negative function ρ, defined for each pair of points x, y in X, subject to the conditions (i) $\rho(x, y) = 0$ if and only if $x = 0$, (ii) $\rho(x, y) = \rho(y, x)$, and (iii) $\rho(x, y) \leqslant \rho(x, z) + \rho(z, y), z \in X$ *(triangle inequality)*. X, endowed with the metric ρ, is called a *metric space*. In a metric space X the set of all points $\{y \mid d(x, y) < \varepsilon\}$ is called the *open ball* of radius ε about x. The *metric topology* for X is the weakest topology whose open sets contain the open balls of X.

Lemma 5. *The set of all open balls about x forms a base for the metric topology at the point x in X.*

Lemma 6. *A metric space is a Hausdorff space.*

If Φ is the field of real or complex numbers and if ρ is defined by $\rho(\alpha, \beta) = |\alpha - \beta|, \alpha, \beta \in \Phi$, then ρ is a metric on Φ and the topology induced by this metric is called the *usual topology* (for Φ).

Theorem 7. *Every one-dimensional Hausdorff linear topological space over the field Φ is topologically isomorphic with Φ in its usual topology. If $0 \neq x_1 \in X$, a topological isomorphism of Φ onto X is defined by $T\alpha = \alpha x_1$, $\alpha \in \Phi$.*

In a metric space X a sequence $\{x_n\}$ *converges* to x in X if and only if $\lim\limits_n d(x_n, x) = 0$. $\{x_n\}$ is a *Cauchy sequence* if and only if $\lim\limits_{m,n} d(x_m, x_n) = 0$, i.e. if for every $\varepsilon > 0$ there is an n_ε such that $d(x_m, x_n) < \varepsilon$ whenever $m, n \geqslant n_\varepsilon$. X is said to be *complete* if every Cauchy sequence is convergent in X. If X is complete, then it is easy to see that every net $\{x_\lambda\}$ in X such that $\lim_{\lambda,\mu} d(x_\lambda, x_\mu) = 0$ (a *Cauchy net*) is convergent in X.

Theorem 8. *If X is a complete linear metric space under each of two metrics, and if one of the corresponding topologies is weaker than the other, then the two topologies are equal.*

A subset A of a metric space is said to be *totally bounded* if for every $\varepsilon > 0$ there is a finite set N of points in A such that for any x in A there is a y in N for which $d(x, y) < \varepsilon$ (i.e. there is a finite ε-net in A).

Lemma 9. *A subset A of a metric space is compact if and only if it is closed and sequentially compact. Moreover, $\bar{A}$ is compact if and only if $\bar{A}$ is complete and A is totally bounded.*

Let X be a linear space. A metric ρ on X is said to be *invariant* if $\rho(x, y) = \rho(x - y, 0)$, $x, y \in X$. Let ρ be an invariant metric on X with the additional properties (i) $\lim_n \rho(\alpha_n x, 0) = 0$ whenever $\{\alpha_n\}$ is a sequence in Φ, x is in X and $\lim_n \alpha_n = 0$ in the usual topology for Φ, and (ii) $\lim_n \rho(\alpha x_n, 0) = 0$ whenever α is in Φ, $\{x_n\}$ is a sequence in X and $\lim_n \rho(x_n, 0) = 0$. Then the real number $\|x\|$, defined by $\|x\| = \rho(x, 0)$, defines the *quasinorm* of an element x in X and X endowed with this quasi-norm is a *quasi-normed linear space*. It follows easily that $\|x\| = 0$ if and only if $x = 0$, $\|x + y\| \leqslant \|x\| + \|y\|$ and $\| - x\| = \|x\|$. An *F-space* is a complete quasi-normed linear space.

Theorem 10. *An F-space is a linear topological space.*

If the quasi-norm in a quasi-normed linear space X satisfies $\|\alpha x\| = |\alpha| \|x\|$, $\alpha \in \Phi$, $x \in X$, it is called a *norm* and X is called a *normed linear space*. The closed set $U = \{x \in X \mid \|x\| \leqslant 1\}$ is called the *unit ball* of X.

Theorem 11. *A normed linear space X is separable if and only if there exists a total sequence in X.*

Theorem 12. *A bounded closed subset of a normed linear space X is compact if and only if X is finite dimensional.*

If Y is a closed linear subspace of the normed linear space X and if we introduce the norm $\|x + Y\| = \inf\{\|y\| \mid y \in x + Y\}$ for each element $x + Y$ of the factor space X/Y we have

Theorem 13. *X/Y is a normed linear space.*

A *Banach space* is a complete normed linear space. The metric topology induced by the norm in a Banach space X is called the *norm* or *strong topology* for X. Since $\|tx+(1-t)y\|\leqslant t\|x\|+(1-t)\|y\|$, $x,y\in X$ and $t\in[0,1]$, the open balls about 0 are convex so that, by Lemma 5, each Banach space X is obviously locally convex.

Theorem 14. *If a linear space X can be made into a Banach space by two different choices of a norm, $\|x\|$ and $\|x\|'$ and if one of them defines a weaker topology than the other, then there exist constants M_1 and M_2 such that $0<M_1\leqslant\|x\|'/\|x\|\leqslant M_2<\infty$ for all $x\neq0$ in X. (i.e. $\|\ \|$ and $\|\ \|'$ are equivalent).*

A *(real or complex) Hilbert space* is a linear space H over the field Φ (of real or complex numbers) together with a Φ-valued function $(\cdot,\cdot)$, called *inner product*, defined for each pair of points x,y in H. This function satisfies the following conditions:

(i) $(x,x)\geqslant0$, $x\in H$; $(x,x)=0$ if and only if $x=0$;

(ii) $(x+y,z)=(x,z)+(y,z)$, $x,y,z\in H$;

(iii) $(\alpha x,y)=\alpha(x,y)$, $\alpha\in\Phi$, $x,y\in H$;

(iv) $(x,y)=\overline{(y,x)}$, $x,y\in H$, and

(v) H is complete with respect to the metric defined by the norm $\|x\|=(x,x)^{\frac{1}{2}}$, $x\in H$.

Theorem 15. *H is a Banach space and for any x,y in H we have $|(x,y)|\leqslant\|x\|\,\|y\|$ (Schwarz's inequality).*

The *orthocomplement* $A^{\perp}$ of a set A in H is the set $A^{\perp}=\{x\in H|(x,A)=0\}$. If A and A' are closed linear subspaces of H such that $A\cap A'=0$ and $A+A'=H$ we write $H=A\oplus A'$ and H is called the *direct sum* of A and A'. Let $A^{\perp\perp}$ denote $(A^{\perp})^{\perp}$.

Lemma 16. *If A is a closed linear subspace of H, then $A^{\perp}$ is also a closed linear subspace of H and we have $H=A\oplus A^{\perp}$. Furthermore $A^{\perp\perp}=A$.*

Let $\delta_{\mu\nu}$ be the *Kronecker symbol*, defined by $\delta_{\mu\nu}=0$, $\mu\neq\nu$; $\delta_{\mu\mu}=1$, where μ and ν are elements of an arbitrary index set. A sequence $\{x_i\}$ in H is said to be *orthonormal* if and only if $(x_i,x_j)=\delta_{ij}$. If $\{x_i\}$ is such an orthonormal sequence, then *Bessel's inequality* applies, i.e.

$$\sum_{i=1}^{\infty}|(x,x_i)|^2\leqslant\|x\|^2 \text{ for every } x \text{ in } H.$$

A special example of a (finite dimensional) Hilbert space is the *Euclidean space E^n*, denoted by $\mathbb{R}^n$ if its field Φ is $\mathbb{R}$ and by $\mathbb{C}^n$ if $\Phi=\mathbb{C}$, the linear space of all n-tuples $\alpha=\{\alpha_1,\ldots,\alpha_n\}$ of numbers in $\mathbb{R}$ (respectively in $\mathbb{C}$) with norm (or length) $\|\alpha\|=(\alpha,\alpha)^{\frac{1}{2}}$ of α in E^n, where the inner product is defined by $(\alpha,\beta)=\sum_{i=1}^{n}\alpha_i\overline{\beta_i}$, $\alpha,\beta\in E^n$. A set $\{\beta_1,\ldots,\beta_n\}$ of n

orthonormal vectors in E^n is called an *orthonormal basis* for E^n and it is a well known fact that such a basis always exists.

Theorem 17. *Each n-dimensional real (complex) Banach space is topologically isomorphic to $\mathbb{R}^n$ (to $\mathbb{C}^n$ respectively).*

A non-empty subalgebra Y of an algebra X is called a *left ideal* of X provided that $xy \in Y$ for all x in X and all y in Y. It is a *right ideal* if the latter condition is replaced by $yx \in Y$ for all x in X and y in Y. If Y is both a left and a right ideal, then it is called a *two-sided ideal*. If $Y \neq X$, Y is called *proper*. Y is *maximal* if it is proper and is not properly contained in any proper ideal of the same type. An algebra X is *commutative* if and only if $xy = yx$, $x, y \in X$. In this case every ideal is two-sided. The *radical* in a commutative algebra X is the intersection of all the maximal ideals in X. X is *semi-simple* if its radical consists of only the vector 0 in X. A *B-algebra* X is a complex Banach space which is also an algebra over the field $\mathbb{C}$, with *unit* e such that $\|e\| = 1$ and $\|xy\| \leqslant \|x\| \|y\|$, $x, y \in X$. An example of a *B*-algebra is (the space) $B(Y)$ (cf. section 2), where Y is a Banach space.

Theorem 18. *Any semi-simple commutative B-algebra has a unique norm topology.*

A *B*-algebra X is (algebraically) *isomorphic* to a *B*-algebra Y if and only if there is an isomorphism T of the linear space X onto the linear space Y such that $T(xx') = T(x)T(x')$, $x, x' \in X$. If one does not request that T is one-to-one, T is called a *homomorphism* of the algebra X onto the algebra Y. X and Y are said to be *topologically isomorphic B*-algebras if T is also a homeomorphism of the Banach space X onto the Banach space Y. If T maps the identity of X onto the identity of Y then T is said to be *proper*.

Theorem 19. *If X is a commutative B-algebra which is isomorphic to a semi-simple commutative B-algebra Y, then X and Y are topologically isomorphic.*

2. Linear Transformations

Let X and Y be linear topological spaces over the same scalar field Φ. A transformation $T: X \to Y$ is said to be *additive* if for all x_1 and x_2 in X we have $T(x_1 + x_2) = Tx_1 + Tx_2$. An additive transformation $T: X \to Y$ is said to be *linear* if $T(\alpha x) = \alpha Tx$, $\alpha \in \Phi$, $x \in X$. The *null-space* of T, denoted by $T^{-1}(0)$, is the set $\{x \in X \mid T(x) = 0\}$. The linear transformation T is *continuous at the point x* in X if for every neighborhood V of Tx

there corresponds a neighborhood U of x with $T(U) \subset V$. T is *continuous* if it is continuous at each point x of X.

Lemma 1. *If T is continuous anywhere, then it is continuous.*

Lemma 2. *T is continuous if and only if $\lim_\lambda Tx_\lambda = T(\lim_\lambda x_\lambda)$ for every convergent net $\{x_\lambda\}$ in X.*

Lemma 3. *If X, Y and Z are linear topological spaces and if both $T_1: X \to Y$ and $T_2: Y \to Z$ are continuous, then the product $T_2 T_1: X \to Z$ is also continuous.*

A family $\{T_\lambda\}$ of linear transformations of X into Y is *equicontinuous* if and only if for each neighborhood V of 0 in Y there is a neighborhood U of 0 in X such that $T_\lambda(U) \subset V$ for each member T_λ in $\{T_\lambda\}$.

Theorem 4. *(Barrel theorem)   Let $\{T_\lambda\}$ be a family of continuous linear transformations of a barrel space to a locally convex space such that the set $\{T_\lambda x\}$ is bounded for each x in X. Then the family is equicontinuous.*

A linear transformation $T': X' \to Y$ is an *extension* of a linear transformation $T: X \to Y$, if $X \subset X'$ and $T'x = Tx$, $x \in X$. Conversely, T is said to be the *restriction* of T'. The linear transformation $I: X \to X$, defined by $Ix = x$, $x \in X$ is called the *identity* of X. If X is a metric space with metric ρ and if Y is a metric space with metric ρ', then the linear map $T: X \to Y$ is said to be *uniformly continuous* if for every $\varepsilon > 0$ there corresponds a $\delta > 0$ such that $\rho(x, x') < \delta$, $x, x' \in X$, implies $\rho'(Tx, Tx') < \varepsilon$. Applied to this situation one has, if Y is complete,

Theorem 5. *If a set D is dense in X and $T: D \to Y$ is uniformly continuous on D, then T has a unique extension $T': X \to Y$ and T' is uniformly continuous on X.*

Theorem 6. *A continuous linear one-to-one transformation of one complete linear metric space onto another has a continuous linear inverse.*

Theorem 7. *A linear transformation of an F-space into another is continuous if and only if it maps bounded sets into bounded sets.*

Theorem 8. *Let $\{T_\lambda\}$ be a net of continuous linear transformations of an F-space X into another F-space Y. If $\lim_\lambda T_\lambda x$ exists for all x in a total subset S of X, and if $\{T_\lambda x\}$ is a bounded set in Y for each x in X, then there exists a continuous linear transformation $T: X \to Y$ such that $Tx = \lim_\lambda T_\lambda x$, $x \in X$.*

Let $T: D \to Y$ be a linear transformation into an F-space Y, with domain D in an F-space X. T is said to be *closed* if, whenever $x_n \in D$,

$x_n \to x$, $Tx_n \to y$, it always follows that $x \in D$ and $Tx = y$. It is clear that T is closed whenever D is closed and T is continuous.

Theorem 9. *If $T: D \to Y$ is closed and if T^{-1} exists, then T^{-1} is also closed.*

Theorem 10. *(Closed graph theorem)   If T is closed and $D = X$, then T is continuous.*

Lemma 11. *Let $T: D \to Y$ be a linear transformation with domain D in a normed linear space X into another, Y. Then T is continuous if and only if T has a finite norm $\|T\|$, defined by $\|T\| = \sup\{\|Tx\| \mid \|x\| \leqslant 1, x \in D\}$.*

In the situation described in the above lemma, T is said to be *bounded.*

Theorem 12. *If T is a bounded linear transformation on $D \subset X$ to Y, then T has a unique bounded linear extension T' on $\bar{D}$ and we have $\|T'\| = \|T\|$.*

An *isometric isomorphism* of a Banach space X onto another, Y, is a topological isomorphism of X onto Y for which $\|Tx\| = \|x\|$, $x \in X$.

Theorem 13. *If T is a bounded linear transformation of a Banach space X onto another, Y, then Y is topologically isomorphic to the factor space $X/T^{-1}(0)$.*

The linear space $B(X, Y)$ $(B(X))$ of all bounded linear transformations of a Banach space X into a Banach space Y (into X respectively) with the definition of the norm which has preceded, is a Banach space. The topology in $B(X, Y)$ defined by this norm is called the *uniform operator topology*. An element of $B(X)$ is called an *endomorphism* of X. Next, we have as a consequence of Theorem 8:

Theorem 14. (BANACH-STEINHAUS) *Let $\{T_n\}$ be a sequence of operators in $B(X, Y)$. If $\lim_n T_n x$ exists for all x in a total subset S of X, and if $\sup_n \|T_n x\| < \infty$ for all x in X, then there exists a T in $B(X, Y)$ such that $Tx = \lim_n T_n x$, $x \in X$ and one has $\|T\| \leqslant \sup_n \|T_n\| < \infty$.*

Theorem 15. *A linear transformation $T: D \to Y$ with domain D in X has a bounded inverse if and only if there is a constant $M > 0$ such that $\|Tx\| \geqslant M\|x\|$ for all x in D. In this case $\|T^{-1}\| \leqslant 1/M$.*

An operator in $B(X, Y)$ is said to be *compact* if it maps the unit ball of X onto a conditionally compact subset of Y.

Theorem 16. *Let $T_1 \in B(X, Y)$ and $T_2 \in B(Y, Z)$. If one of the operators T_1 or T_2 is compact, then the product $T_2 T_1$ in $B(X, Z)$ is also compact.*

Theorem 17. *The set of compact operators in $B(X, Y)$ is closed in the uniform operator topology of $B(X, Y)$.*

A linear transformation $P: X \rightarrow X$ is called a *projection* in X (or of X on $P(X)$, or of X onto $P(X)$) if $P^2 = P$. We call attention to the fact that the definition of P does not necessarily imply that P is in $B(X)$. In this connection, we must know that X is the *direct sum* $M \oplus N$ of two linear subspaces M and N if and only if $M + N = X$ and $M \cap N = 0$, in other words, if and only if every element x in X can be written uniquely as a sum $x = x_1 + x_2$, where x_1 is in M and x_2 is in N.

Theorem 18. *If P is a projection in X, then $X = P(X) \oplus (I - P)(X)$. Conversely, if M and N are (closed) linear subspaces of X such that $X = M \oplus N$, and if $P: X \rightarrow X$ is defined by $Px = x_1$, $x \in X$, where x_1 is the unique component of x in M, then P is a (continuous) projection of X onto M.*

Theorem 19. *If M is a linear subspace of X, there exists a projection of X on M.*

We observe that there may be more than one projection of X onto M. Next, let X be a complex Banach space. The *spectrum* $\sigma(A)$ of an operator $A \in B(X)$ is the complement in $\mathbb{C}$ of the set of all points λ of $\mathbb{C}$ for which $(\lambda I - A)^{-1}$ exists and is in $B(X)$. $R(\lambda, A) = (\lambda I - A)^{-1}$ is called the *resolvent* of A.

Theorem 20. *$\sigma(A)$ is a non-empty compact subset of $\{ \lambda \mid |\lambda| \leqslant \|A\|, \lambda \in \mathbb{C} \}$.*

If $\lambda \in \sigma(A)$ is such that $\lambda I - A$ is not one-to-one, then λ is called an *eigenvalue* of A, in this case there exists an $x \neq 0$ such that $Ax = \lambda x$.

3. Conjugate Spaces and Weak Topologies

Let X be a linear topological space over the field Φ. The *conjugate space* X^* of X is the set of all continuous linear functions of X into Φ and the elements of X^* are called *continuous linear functionals*. A subset A of X^* is said to be *total over X* if for any x in X, $x^*(x) = 0$ for all x^* in A implies that $x = 0$.

Theorem 1. *If X is locally convex and Hausdorff, then X^* is total over X.*

The *weak topology* for a linear topological space X is the topology for X obtained by taking as a base at the points x of X the neighborhoods

$$N(x, \Gamma, \varepsilon) = \{ y \mid |x^*(y - x)| < \varepsilon, x^* \in \Gamma \},$$

where $\varepsilon>0$ and Γ is a finite subset of X^*. If X is a Banach space, the weak topology for X is, of course, weaker than the initial (i.e. norm) topology for X.

Theorem 2. *The weak topology for X is locally convex. It is Hausdorff whenever the topology for X is locally convex and Hausdorff.*

Theorem 3. *x^* belongs to X^* if and only if x^* is continuous in the weak topology for X.*

A set X is said to be *weakly closed (weakly bounded)* if and only if it is closed (bounded) in the weak topology for X.

Theorem 4. *If X is locally convex, then a convex subset of X is closed if and only if it is weakly closed, and a subset of X is bounded if and only if it is weakly bounded.*

Theorem 5. *Let X be defined as before and let Y be a linear subspace of X. Then to every continuous linear functional f on Y there exists an $x^*\in X^*$ such that $x^*(y)=f(y), y\in Y$.*

Theorem 6. *Let X be locally convex, let Y be a closed linear subspace of X and suppose that there exists an x in X which is not in Y. Then there exists an x^* in X^* such that $x^*(y)=0$, $y\in Y$ and $x^*(x)=1$.*

Let now X be a Banach space over the field Φ and let X^* be the conjugate space of X endowed with the norm defined by $\|x^*\|=\sup\{|x^*(x)|\,|\,\|x\|\leqslant1\}$, $x^*\in X^*$. The set $Y^{\perp}=\{x^*\in X^*|x^*(Y)=0\}$ is called the *orthogonal complement* of the linear subspace Y of X.

Lemma 7. *X^* is a Banach space.*

Theorem 8. *Let Y be a closed linear subspace of X. Then Y^* is isometrically isomorphic to the factor space $X^*/Y^{\perp}$.*

It is clear that X^* itself possesses a conjugate (Banach) space, X^{**}, the second conjugate space and we denote the sequence of spaces obtained by continued conjugation by $X, X^*, X^{**}, X^{***}, \dots$. We have the following important theorem on the existence of extensions of continuous linear functionals:

Theorem 9. (HAHN-BANACH) *If Y is a linear subspace of X, then to every y^* in Y^* there exists an x^* in X^* such that $x^*(y)=y^*(y)$, $y\in Y$ and $\|x^*\|=\|y^*\|$.*

The next theorem shows that for every $x\neq y$ in X there is at least one x^* in X^* such that $x^*(x)\neq x^*(y)$. In other words there are enough elements in X^* to distinguish between points of X:

Theorem 10. *To each* $x \neq 0$ *in* X *there exists an* x^* *in* X^* *with* $\|x^*\| = 1$ *and* $x^*(x) = \|x\|$. *Therefore,* $\|x\| = \sup\{|x^*(x)| \mid \|x^*\| \leqslant 1\}$ *for every* x *in* X.

A *determining manifold* for X is a closed linear subspace Γ of X^* such that $\|x\| = \sup\{|x^*(x)| \mid \|x^*\| \leqslant 1, x^* \in \Gamma\}$. Obviously, X^* itself is a determining manifold for X.

Theorem 11. *If* X^* *is separable, so is* X.

However, the conjugate space of a separable Banach space X need not be separable. An example is $X = l_1$ (cf. section I.4b) where l_1^* is isometrically isomorphic to the non-separable space l_∞. The linear transformation $J : X \to X^{**}$ defined by $Jx(x^*) = x^*(x)$, $x^* \in X^*$ is called the *natural embedding* of X into X^{**}.

Theorem 12. J *is an isometric isomorphism of* X *into* X^{**}.

Let Λ be an arbitrary index set. Then the following two theorems are consequences of the *principle of uniform boundedness*.

Theorem 13. *Let* $\{x_\lambda \mid \lambda \in \Lambda\}$ *be an indexed set in* X. *Then* $\sup\{\|x_\lambda\| \mid \lambda \in \Lambda\} < \infty$, *whenever* $\sup\{|x^*(x_\lambda)| \mid \lambda \in \Lambda\} < \infty$ *for all* x^* *in* X^*.

Theorem 14. *Let* $\{T_\lambda \mid \lambda \in \Lambda\}$ *be a family of bounded linear transformations of* X *into a Banach space* Y. *Then* $\sup\{\|T_\lambda\| \mid \lambda \in \Lambda\} < \infty$ *if and only if* $\sup\{|y^*(T_\lambda x)| \mid \lambda \in \Lambda\} < \infty$, $x \in X$, $y^* \in Y^*$.

A net $\{x_\lambda\}$ in X is said to be *weakly convergent* if $\lim_\lambda x^*(x_\lambda)$ exists for each x^* in X^*, it is said *weakly convergent to a point* x *in* X if $\lim_\lambda x^*(x_\lambda) = x^*(x)$ for every x^* in X^*. $\{x_\lambda\}$ converges to x in the weak topology if and only if it converges weakly to x. A subset A of X is said to be *(conditionally) weakly sequentially complete* if every weakly convergent sequence in A converges weakly (i.e. in the weak topology for X) to an element (of X) of A.

Theorem 15. *If* $\{x_n\}$ *is a weakly convergent sequence in* X, *then* $\sup_n \|x_n\| < \infty$. *If it converges weakly to* x *in* X, *then* $x \in \overline{\mathrm{sp}}\{x_n\}$ *and* $\|x\| \leqslant \sup_n \|x_n\|$.

We remember that the weak topology for X is weaker than the norm topology for X (defined by the neighborhoods $N(x,\varepsilon) = \{y \mid \|y - x\| < \varepsilon\}$ of the points x in X). It is known that the weak topology for X is a metric topology if and only if X^* is separable.

Theorem 16. *Let* S *be a total set in* X^* *and let* $\{x_n\}$ *be a sequence in* X *such that* $\sup_n \|x_n\| < \infty$ *and for some* x *in* X, $\lim_n y^*(x_n) = y^*(x)$, $y^* \in S$. *Then* $\{x_n\}$ *converges to* x *in the weak topology for* X.

X is *reflexive* if the natural embedding $J:X \to X^{**}$ is onto. Finite dimensional vector spaces are examples of reflexive spaces.

Theorem 17. *If X is reflexive, it is weakly sequentially complete.*

Theorem 18. *The following statements are equivalent:*
(i) *X is reflexive.*
(ii) *X is topologically isomorphic to a reflexive Banach space.*
(iii) *X^* is reflexive.*
(iv) *The unit ball U of X is sequentially compact in the weak topology for X.*
(v) *Every closed linear subspace of X is reflexive.*
(vi) *Each separable closed linear subspace of X is reflexive.*

Theorem 19. *If X^* is separable and the unit ball U of X is weakly sequentially complete, then X is reflexive.*

The *weak* topology* for the conjugate space X^* can be introduced by taking as a base at the points x^* of X^* the neighborhoods

$$N(x^*, \Gamma, \varepsilon) = \{ y^* \,|\, |(y^* - x^*)(x)| < \varepsilon, \ x \in \Gamma \},$$

where $\varepsilon > 0$ and Γ is a finite subset of X. A net $\{x_\lambda^*\}$ in X^* converges to x^* in the weak* topology for X^* if and only if $\lim_\lambda x_\lambda^*(x) = x^*(x)$ for all x in X. The weak* topology for X^* is weaker than the weak topology for X^*. The weak* topology for X^* is a locally convex topology. It is a metric topology on bounded sets if and only if X is separable.

Theorem 20. (BANACH) *A linear functional f on X^* is of the form $f(x^*) = x^*(x), x^* \in X^*$, with a certain x in X depending on f, if and only if f is continuous in the weak* topology for X^*.*

Theorem 21. (ALAOGLU) *The unit ball U^* of X^* is compact in the weak* topology for X^*. If X is separable the compactness of U^* is sequential.*

Theorem 22. *Let J be the natural embedding of X into X^{**} and let U, U^{**} be the unit balls in X, X^{**} respectively. Then $J(U)$ is weakly* dense in U^{**} and $J(X)$ is weakly* dense in X^{**}.*

The *adjoint* of a bounded linear operator T of X into a Banach space Y is the mapping $T^*:Y^* \to X^*$, defined by $T^*y^*(x) = y^*(Tx)$, $y^* \in Y^*$, $x \in X$. Of course, T^* is linear.

Theorem 23. $\|T^*\| = \|T\|$ *and I^* is the identity in $B(X^*)$.*

Theorem 24. *If T is onto, then T^* is a topological isomorphism of Y^* onto $X^* \cap T^{-1}(0)^{\perp}$.*

Theorem 25. *If T is a topological isomorphism of X onto Y, then T^* is a topological isomorphism of Y^* onto X^*. In this case one has $(T^{-1})^* = (T^*)^{-1}$.*

Theorem 26. *If S is in $B(X, Y)$, and T is in $B(Y, Z)$, where X, Y and Z are all Banach spaces, then $(S\,T)^* = T^*\,S^*$.*

Theorem 27. T^* *is compact if and only if T is compact.*

Theorem 28. *If P is a continuous projection in X, then P^* is likewise a continuous projection in X^* and $P^*(X^*) = (I - P)(X)^{\perp}, (I^* - P^*)(X^*) = P(X)^{\perp}$.*

Let H be a Hilbert space and T an operator in $B(H)$. There is a unique operator T^* in $B(H)$, called the *Hilbert space adjoint* of T, which is defined by $(Tx, y) = (x, T^* y)$, $x, y \in H$. T is called *self-adjoint* if $T^* = T$.

Theorem 29. *Let $P \neq 0$ be a projection in H. Then $(I - P)(H) = P(H)^{\perp}$ if and only if $\|P\| = 1$.*

A projection with the properties written in the foregoing theorem is said to be *orthogonal*.

Theorem 30. *A projection in H is orthogonal if and only if it is self-adjoint.*

4. Special Banach Spaces

At this place we write down some properties of a few special Banach spaces and some theorems on these spaces which are of interest in the theory and application of bases.

Let s be the linear space (over the field Φ of real or complex numbers) of all sequences $\alpha = \{\alpha_n\}$ in Φ.

a) The set c in s of all convergent sequences, endowed with the norm $\|\alpha\| = \sup_n |\alpha_n|$ is a separable Banach space. c_0 is the closed linear subspace of c of all sequences converging to zero. Neither c nor c_0 are weakly sequentially complete and both, c and c_0, are not reflexive. c^* and c_0^* are both isometrically isomorphic to the space l_1 which is described under b).

Theorem 1. (PHILLIPS) *If J is the natural embedding of c_0 into c_0^{**} and if $\{x_n\}$ is a sequence in c_0^{***} which converges to zero in the weak* topology for c_0^{***}, then $\lim_n \|J^* x_n\| = 0$.*

b) The set l_p, $1 \leqslant p < \infty$ of all elements α of s with finite norm

$$\|\alpha\| = \left(\sum_{n=1}^{\infty} |\alpha_n|^p \right)^{1/p}$$ is a separable weakly sequentially complete Banach

space. l_1 is not reflexive. If $1 < p < \infty$ and $1/p + 1/q = 1$, l_p is reflexive and l_p^* is isometrically isomorphic with l_q, where the corresponding

isometric isomorphism T is given by $T\beta(\alpha) = \sum_{i=1}^{\infty} \alpha_i \beta_i, \alpha \in l_p, \beta \in l_q$.

Taking $\alpha \in c_0$ and $\beta \in l_1$ in the preceding definition, T defines an iso-metric isomorphism of l_1 onto c_0^*. It is not misleading, therefore, to identify sometimes c_0^* with l_1 or l_p^* with l_q.

For $p = 1$ we have the following important theorem:

Theorem 2. *In l_1 weak and strong convergence of sequences are the same.*

In the special case $p = 2$, l_p is a Hilbert space (with inner product

$$(\alpha, \beta) = \sum_{n=1}^{\infty} \alpha_n \bar{\beta}_n).$$

c) The set l_∞ of all bounded sequences α in Φ with norm $\|\alpha\| = \sup_n |\alpha_n|$

is a Banach space which is neither separable nor weakly sequentially complete. It is easy to see that the set of all characteristic functions of the subsets of the set of positive integers forms a total set in l_∞. The

mapping $T: l_\infty \to l_1^*$, defined by $T\beta(\alpha) = \sum_{n=1}^{\infty} \alpha_n \beta_n, \alpha \in l_1, \beta \in l_\infty$, is an

isometric isomorphism of l_∞ onto l_1^*. It is known that weak and weak* convergence of sequences are equivalent in l_∞^*. Furthermore, every bounded linear transformation of l_∞ into a weakly sequentially complete Banach space sends weak Cauchy sequences into strongly convergent sequences.

Theorem 3. *Every separable Banach space is isometrically isomorphic to a closed linear subspace of l_∞.*

d) Let S be a compact metric space and let X be a Banach space. The linear space $C(S, X)$ over the field Φ, of all continuous ($=$ uniformly continuous) vector valued functions $f: S \to X$ for which the norm $\|f\| = \sup \{\|f(s)\| \,|\, s \in S\}$ is finite is a separable Banach space, where $\|f(s)\|$ is the norm of $f(s)$ as an element of X. If $X = \Phi$ we simply write $C(S)$. We recall that a set E in $C(S, X)$ is equicontinuous if and only if to every $\varepsilon > 0$ and every s in S there is a neighborhood N of s in S (which may depend on s) such that $\sup \{\|f(s) - f(t)\| \,|\, t \in N\} < \varepsilon$ for all f in E.

Theorem 4. *Let* $\{f_n\}$ *be an equicontinuous sequence in* $C(S,X)$. *If* f_n *converges pointwise (i.e. for each point s in S) in X to a function f in* $C(S,X)$, *then* f_n *converges uniformly (on S) to f.*

Theorem 5. (STONE-WEIERSTRASS) *Let A be a closed subalgebra of* $C(S)$ *which contains the characteristic function* χ_S *of S (which is the unit in* $C(S)$ *) and which contains* $\bar{A} = \{\overline{f(\cdot)} \,|\, f \in A\}$, *if* $\Phi = \mathbb{C}$. *Moreover, if for every pair s, t of distinct points in S there is an f in A with* $f(s) \neq f(t)$, *then A is dense in* $C(S)$.

As a special instance we take $S = [0,1]$ and $X = \Phi$ and we write for the corresponding space $C[0,1]$. It is known that $C[0,1]$ is not weakly sequentially complete. However $C^*[0,1]$ is weakly sequentially complete, but not separable.

e) Let S be a compact interval in $\mathbb{R}$ and let p be a real number with $1 \leqslant p < \infty$. The linear space $L_p(S)$ is the set of all equivalence classes of functions $f: S \to \Phi$ which are measurable and p th-power integrable on S. $L_p(S)$ is a separable and weakly sequentially complete Banach space with norm $\|f\|_p = \left[\int_S |f(s)|^p \, ds \right]^{1/p}$. Hence, if f and g are in $L_p(S)$, then the sum $f + g$ is in $L_p(S)$ and $\|f+g\|_p \leqslant \|f\|_p + \|g\|_p$ which is *Minkowski's inequality.*

Theorem 6. *If* $1 < p < \infty$ *and* $1/p + 1/q = 1$, $L_p(S)$ *is reflexive and* $L_p^*(S)$ *is isometrically isomorphic with* $L_q(S)$; *the corresponding isometric isomorphism* $T: L_q(S) \to L_p^*(S)$ *is given by* $Tg(f) = \int_S f(s)g(s) \, ds, f \in L_p(S)$, $g \in L_q(S)$.

Furthermore, if f is in $L_p(S)$ and g is in $L_q(S)$, then the product fg is in $L_1(S)$ and $\|fg\|_1 \leqslant \|f\|_p \|g\|_q$ which is *Hölder's inequality.* $L_2(S)$ is a Hilbert space (with inner product $(f,g) = \int_S f(s)\overline{g(s)} \, ds$, $f, g \in L_2(S)$). Very useful is the following theorem on the possibility of interchanging the order of integration in a double integral.

Theorem 7. (FUBINI-TONELLI) *Let S and T be compact intervals in* $\mathbb{R}$ *and let* $f(s,t)$ *be a* Φ-*valued measurable function on* $S \times T$ *such that* $f(s, \cdot)$ *is in* $L_1(T)$ *almost everywhere in S and such that* $\int_T f(s,t) \, dt$ *is in* $L_1(S)$. *Then*

$$\int_S \left(\int_T f(s,t) \, dt \right) ds = \int_T \left(\int_S f(s,t) \, ds \right) dt .$$

Theorem 8. *For* $1 \leqslant p < \infty$ *the subset* $C(S)$ *of* $L_p(S)$ *is dense in* $L_p(S)$.

Theorem 9. *(Lebesgue dominated convergence theorem)* *If g is in* $L_1(S)$ *and if* $\{f_n\}$ *is a sequence in* $L_1(S)$ *such that* $|f_n(s)| \leqslant |g(s)|$ *almost*

everywhere in S, which converges almost everywhere to a function f on
S. Then f is in $L_1(S)$ and we have $\lim\limits_{n} \int\limits_{S} f_n(s)\,ds = \int\limits_{S} f(s)\,ds$.

Theorem 10. *If $\{f_n\}$ is a monotone increasing sequence of non-negative real measurable functions on S, converging almost everywhere to a function f. Then* $\lim\limits_{n} \int\limits_{S} f_n(s)\,ds = \int\limits_{S} f(s)\,ds$.

Next, let $\{x_n\}$ be the trigonometrical system, given by $x_0(s)=(2\pi)^{-\frac{1}{2}}$, $x_{2n-1}(s)=\pi^{-\frac{1}{2}}\sin(ns)$, $x_{2n}(s)=\pi^{-\frac{1}{2}}\cos(ns)$, $n=1,2,\ldots,s\in[0,2\pi]$.

Theorem 11. *For each x in $L_p[0,2\pi]$, $1<p<\infty$, the series* $\sum\limits_{n=0}^{\infty} \int\limits_{0}^{2\pi} x_n(s)x(s)\,ds\cdot x_n$ *converges to x in the norm topology for $L_p[0,2\pi]$.*

Corollary 12. *The sequence $\{x_n\}$ is total in $L_p[0,2\pi]$, $1<p<\infty$.*

f) Let D be the open unit disc $\{z\,\|z\|<1\}$ in the complex plane $\mathbb{C}$. The linear space $A(D)$ of all complex valued functions f which are analytic on D and continuous on $\bar{D}$ is a Banach space if the norm in $A(D)$ is defined by $\|f\|=\sup\{|f(z)|\,|z\in D\}$. Let z_0 be an arbitrary point in D. Then every function f in $A(D)$ has the Taylor series expansion $f(z) = \sum\limits_{n=0}^{\infty} (f^{(n)}(z_0)/n!)(z-z_0)^n$ which converges absolutely and uniformly for z in any closed disc $\{z\,|z-z_0|\leqslant r\}$ of radius r which is contained in D. The Taylor series expansion is unique since we have

Theorem 13. *If two power series $\sum\limits_{n=0}^{\infty} \alpha_n(z-z_0)^n$ and $\sum\limits_{n=0}^{\infty} \beta_n(z-z_0)^n$ converge on a neighborhood of z_0 absolutely to the same sum, then $\alpha_n=\beta_n$ for all n.*

g) Let D be defined as under f) and let μ be the planar Lebesgue measure in D. Let A^2 be the linear space of all complex valued functions on D which are analytic in D and square-integrable with respect to μ. Defining the inner product (f,g) of f and g in A^2 by $(f,g)=\int\limits_{D} f(z)\overline{g(z)}\,d\mu(z)$, A^2 becomes a Hilbert space.

References for Chapter I: Bourbaki [1], Day [2], Dieudonné [2], Dunford and Schwartz [1], Edwards [2], Halmos [1 and 2], Hausdorff [1], Hille and Phillips [1], Kelley and Namioka [1], Köthe [2], Rickart [1], Taylor [4], Wilansky [2], Yosida [1] and Zygmund [1].

CHAPTER II

Convergence of Series in Banach Spaces

It is essential for understanding the concepts of bases of different types to have some knowledge about the various definitions of convergence of a series and to stress its hierarchy and its interconnections. This is done in the first paragraph. There we also present the proof of the important Orlicz-Pettis theorem which claims the equivalence of weak and strong unconditional convergence of series in Banach spaces. The second paragraph contains Riemann's theorem which asserts that absolute and unconditional convergence of series in finite dimensional vector spaces are the same, and the famous Dvoretzky-Rogers theorem. The latter states the existence in every infinite dimensional Banach space of an unconditional series which is not absolutely convergent, a fact, which has been conjectured for about twenty years and which has been settled down by DVORETZKY and ROGERS in 1950.

1. Relations among Different Types of Convergence

There are several different definitions for unconditional convergence of a series in a Banach space X. However, it can be shown that all forms defined in the following definition are equivalent (see HILDEBRANDT [1], ORLICZ [2] and DAY [1]). It is assumed that in each case the series $\sum_i x_i$ is convergent in X and we take $x = \sum_i x_i$.

Definition 1.

(i) $\sum_i x_i$ is *unconditionally (or reordered) convergent if for every permutation p of the integers the series $\sum_i x_{p(i)}$ converges in X.*

(ii) $\sum_i x_i$ is *unordered convergent if* $\lim_{\sigma \in \Sigma} \sum_{i \in \sigma} x_i = x$, *where Σ is the set of all finite subsets of the set of integers, directed by $\supseteq$.*

(iii) $\sum_i x_i$ is subseries convergent if for every increasing sequence $\{n_i\}$ of integers the series $\sum_i x_{n_i}$ converges to some element of X.

(iv) $\sum_i x_i$ is bounded multiplier convergent if for each bounded sequence $\{\alpha_i\}$ in Φ, the series $\sum_i \alpha_i x_i$ is convergent to some element of X.

We note that the convergence in all the preceding definitions is the strong convergence in X. Before showing the equivalence of (i) to (iv) we prove the famous Orlicz-Pettis theorem which links subseries convergence of series in the strong and weak topologies of X.

Theorem 2. (ORLICZ-PETTIS) *If a series is subseries convergent in the weak topology of X, then it is subseries convergent in the strong topology of X.*

Proof. Let the series of the theorem be $\sum_i x_i$ and let Σ be the set of all increasing sequences μ of positive integers. By assumption there exists for each μ in Σ an x_μ in X such that

$$\sum_{i \in \mu} x^*(x_i) = x^*(x_\mu)$$

for every x^* in X^*. Since subseries convergence in Φ is equivalent to absolute convergence (cf. to Definition 2.1 and to Theorem 2.2 for the more general case), it follows that $\sum_{i=1}^{\infty} |x^*(x_i)| < \infty$ for all x^* in X^*. Now, $T:X^* \to l_1$, defined by $Tx^* = \{x^*(x_i)\}$ is clearly linear. T is closed since from $\lim_n x_n^* = x^*$, $\{x_n^*\} \subset X^*$, and from the existence of $\lim_n T x_n^*$ in l_1 it follows $\lim_n T x_n^* = \lim_n \{x_n^*(x_i)\} = \{\lim_n x_n^*(x_i)\} = \{x^*(x_i)\} = Tx^*$. But T, being closed and defined everywhere on X^*, must be bounded (I.2.10). Since x_μ is the weak limit of a sequence in X, we have $x_\mu \in \overline{sp}\{x_i\}$ (I.3.15). Let now $\{x_n^*\}$ be a sequence in the closed unit ball U^* of X^*, and let $\{y_n^*\}$ be its restriction on $Y = \overline{sp}\{x_i\}$. Since Y is separable, by Theorem I.3.21, the unit ball in Y^* is sequentially compact in the weak* topology of Y^*. Thus there exists a subsequence $\{y_{n_j}^*\}$ of $\{y_n^*\}$, converging to an element y_0^* in Y^* in the weak* topology of Y^*. As a consequence of the Hahn-Banach theorem there is a continuous extension x_0^* of y_0^* to the whole space X. Let χ_μ be the characteristic function of μ. Evidently, χ_μ is in l_∞ and the corresponding functional f_μ, defined by $f_\mu(\alpha) = \sum_i \chi_\mu(i)\alpha_i = \sum_{i \in \mu} \alpha_i$ for every $\alpha = \{\alpha_i\}$ in l_1, is an element of l_1^*. It then follows

$$\lim_j f_\mu(Tx_{n_j}^*) = \lim_j \sum_{i\in\mu} x_{n_j}^*(x_i) = \lim_j x_{n_j}^*(x_\mu) = \lim_j y_{n_j}^*(x_\mu)$$

$$= y_0^*(x_\mu) = x_0^*(x_\mu) = \sum_{i\in\mu} x_0^*(x_i) = f_\mu(Tx_0^*).$$

Now by I.4.c, $\{\chi_\mu | \mu \in \Sigma\}$ is a total set in l_∞ and since the mapping which relates χ_μ and f_μ is an isometric isomorphism of l_∞ onto l_1^*, $\{f_\mu | \mu \in \Sigma\}$ is a total set in l_1^*. Hence, using the fact that the boundedness of T implies that of the sequence Tx_n^*, $Tx_n^*,\ldots$, we conclude from Theorem I.3.16, that $Tx_{n_j}^*$ converges to Tx_0^* in the weak topology of l_1. Since weak and strong convergence in l_1 is the same (I.4.2) it is clear that $\overline{T(U^*)}$ is compact (I.1.9).

Next, let $\{T_k\}$ be a sequence of operators of l_1 to itself defined by $T_k\alpha = \{0,\ldots,0,\alpha_k,\alpha_{k+1},\ldots\}$. Obviously, $\|T_k\| = 1$ for all k so that $\{T_k\}$ is an equicontinuous set of functions on the compact space $\overline{T(U^*)}$ to l_1. Since $\lim_k T_k\alpha = 0$ for each α in $\overline{T(U^*)}$ it follows from Theorem I.4.4 that $\limsup_k \{\|T_k\alpha\| \, | \, \alpha \in T(U^*)\} = 0$, hence that

$$\lim_k \sum_{i=k}^\infty |x^*(x_i)| = 0 \qquad (1)$$

uniformly on U^*. Finally,

$$\left\|x_\mu - \sum_{\substack{i\in\mu \\ i<k}} x_i\right\| \leqslant \sup\left\{\sum_{\substack{i\in\mu \\ i\geqslant k}} |x^*(x_i)| \, \Big| \, x^*\in U^*\right\}$$

$$\leqslant \sup\left\{\sum_{i=k}^\infty |x^*(x_i)| \, \Big| \, x^*\in U^*\right\}.$$

Because the last term converges to zero for $k\to\infty$, $\sum_i x_i$ is strongly subseries convergent and the proof of the theorem is finished.

Theorem 3. *The conditions* (i), (ii), (iii) *and* (iv) *of Definition 1 are equivalent conditions for unconditional convergence of a series* $\sum_i x_i$ *in* X.

Proof. (i) $\Rightarrow$ (ii). Let the series be reordered convergent and suppose that $\lim_{\substack{\sigma\in\Sigma \\ i\in\sigma}} \sum x_i = x$ is not true. Then there exists an $\varepsilon > 0$ such that for all $\sigma\in\Sigma$ there is a $\sigma'\supseteq\sigma$ with $\left\|x - \sum_{i\in\sigma'} x_i\right\| \geqslant \varepsilon$. Since $\sum_i x_i$ is convergent we can choose an n_0 such that $\left\|x - \sum_{i\leqslant n} x_i\right\| < \varepsilon/2$ for $n\geqslant n_0$. We define the sequence $\sigma_1\subseteq\sigma_2\subseteq\cdots\subseteq\sigma_n\subseteq\ldots$ in Σ such that the odd numbered sets σ_n contain all the integers less than or equal to $\max[n_0, \sup\{i | i\in\sigma_n\}]$ and the even numbered sets σ_n satisfy $\left\|x - \sum_{i\in\sigma_n} x_i\right\| \geqslant \varepsilon$. It is now apparent that $\left\| \sum_{i\in\sigma_n} x_i - \sum_{i\in\sigma_{n-1}} x_i \right\| \geqslant \left| \left\| x - \sum_{i\in\sigma_n} x_i \right\| - \left\| x - \sum_{i\in\sigma_{n-1}} x_i \right\| \right| > \varepsilon - \varepsilon/2$,   which

contradicts the hypothesis that $\sum_i x_i$ is reordered convergent. Consequently, $\lim\limits_{\sigma\in\Sigma} \sum_{i\in\sigma} x_i = x$.

(ii) $\Rightarrow$ (i). Let $\lim\limits_{\sigma\in\Sigma} \sum_{i\in\sigma} x_i = x$. Then for every $\varepsilon > 0$ there exists a σ_ε such

that $\left\| x - \sum_{i\in\sigma} x_i \right\| < \varepsilon$ for all $\sigma \supseteq \sigma_\varepsilon$. If p is any permutation of the (positive)

integers, there exists an index n_0 such that $\sigma_\varepsilon \subseteq \{p(1),\ldots,p(n_0)\}$. Hence
for $n \geqslant n_0$ we have $\left\| x - \sum_{i\leqslant n} x_{p(i)} \right\| < \varepsilon$, showing that $\sum_i x_{p(i)} = x$ for every p.

(ii) $\Rightarrow$ (iii). If $x = \lim\limits_{\sigma\in\Sigma} \sum_{i\in\sigma} x_i$ exists, then for every $\varepsilon > 0$ there is an

element σ_ε in Σ for which $\left\| x - \sum_{i\in\sigma} x_i \right\| < \varepsilon$ if $\sigma \supseteq \sigma_\varepsilon$. Let $\{n_i\}$ be any in-

creasing sequence of integers and let $\sigma_p = \sigma_\varepsilon \bigcup \{n_1,\ldots,n_p\}$ for positive
integers p. Then

$$\left\| \sum_{i=p}^{q} x_{n_i} \right\| = \left\| \sum_{i\in\sigma_q} x_i - \sum_{i\in\sigma_{p-1}} x_i \right\| \leqslant \left\| x - \sum_{i\in\sigma_q} x_i \right\| + \left\| x - \sum_{i\in\sigma_{p-1}} x_i \right\| < 2\varepsilon$$

for every $q \geqslant p$ and p such that $n_{p-1} \geqslant \sup\{i | i \in \sigma_\varepsilon\}$. Therefore $\left\{ \sum\limits_{i\leqslant n} x_{n_i} \right\}$

is a Cauchy sequence and the completeness of X implies the convergence
of $\sum_i x_{n_i}$.

(iii) $\Rightarrow$ (ii). Suppose (iii) is true, but that $\lim\limits_{\sigma\in\Sigma} \sum_{i\in\sigma} x_i$ does not exist.

Then there exists an $\varepsilon > 0$ and sequences σ_k^+ and σ_k^- of mutually
disjoint sets in Σ such that $\inf\limits_k \left\| \sum_{i\in\sigma_k^+} x_i - \sum_{i\in\sigma_k^-} x_i \right\| \geqslant 2\varepsilon$. Thus by $\left\| \sum_{i\in\sigma_k^+} x_i \right\|$

$+ \left\| \sum_{i\in\sigma_k^-} x_i \right\| \geqslant \left\| \sum_{i\in\sigma_k^+} x_i - \sum_{i\in\sigma_k^-} x_i \right\| \geqslant 2\varepsilon$, there is a sequence $\{\sigma_k\}$ of mutually

disjoint sets in Σ (with $\sigma_k = \sigma_k^+$ or σ_k^-) such that $\inf\limits_k \left\| \sum_{i\in\sigma_k} x_i \right\| \geqslant \varepsilon$. Now,

let $\{\tau_k\}$ be a subsequence of $\{\sigma_k\}$ such that $\sup\{i | i \in \tau_k\} < \inf\{i | i \in \tau_{k+1}\}$
for all k. Then it is clear that a subseries of $\sum_i x_i, \sum_k \sum_{i\in\tau_k} x_i$ is not Cauchy.

Since X is complete this contradiction shows that $\lim\limits_{\sigma\in\Sigma} \sum_{i\in\sigma} x_i$ exists and
the limit must obviously be x.

(iii) $\Rightarrow$ (iv). Let $\sum_i x_i$ be subseries convergent. Then by (1) there is

each $\varepsilon > 0$ an index n such that for all $k \geqslant n$,

$$\sup\left\{ \sum_{i=k}^{\infty} |x^*(x_i)| \,\Big|\, x^* \in X^*, \|x^*\| \leqslant 1 \right\} < \varepsilon.$$

Given a bounded sequence $\{\alpha_i\}$ in Φ we thus have for every $p, q \geqslant n$,

$$\left\| \sum_{i=p}^{q} \alpha_i x_i \right\| = \sup \left\{ \left| \sum_{i=p}^{q} \alpha_i x^*(x_i) \right| \, \Big| \, \|x^*\| \leqslant 1 \right\}$$

$$\leqslant \sup_i |\alpha_i| \sup \left\{ \sum_{i=p}^{q} |x^*(x_i)| \, \Big| \, \|x^*\| \leqslant 1 \right\}$$

$$< 2\varepsilon \sup_i |\alpha_i|.$$

(iv) then follows as an immediate consequence of the completeness of X.

(iv) $\Rightarrow$ (iii). Suppose $\sum_i x_i$ is bounded multiplier convergent. Let $\{x_{n_i}\}$ be an arbitrary subsequence of $\{x_i\}$. To obtain (iii) we have only to take a multiplying sequence $\{\alpha_j\}$, given by $\alpha_j = 1$ if j is contained in the set $\{n_1, n_2, ...\}$ and $\alpha_j = 0$ else. Finally, this completes the proof of the theorem.

The result at the end of the proof (ii) $\Rightarrow$ (i) suggests the following

Corollary 4. *If the series $\sum_i x_i$ is unconditionally convergent then* $\sum_i x_{p(i)} = \sum_i x_i$ *for every permutation p of the integers.*

2. Unconditional and Absolute Convergence

Definition 1. *A series $\sum_{i=1}^{\infty} x_i$ in a Banach space X is absolutely convergent if and only if $\sum_{i=1}^{\infty} \|x_i\|$ is convergent.*

It is clear that every absolutely convergent series is unconditionally convergent. However, the converse is not true in an infinite dimensional Banach space as we see in the sequel.

Theorem 2. (RIEMANN) *In a finite dimensional Banach space X, unconditional convergence of a series is equivalent to absolute convergence of the series.*

Proof. Since an n-dimensional Banach space X over the field $\mathbb{R}$ (or $\mathbb{C}$) is topologically isomorphic to the Euclidean space $\mathbb{R}^n$ (respectively $\mathbb{C}^n$) (I.1.17), it is sufficient to show that every unconditionally convergent series $\sum_{i=1}^{\infty} \alpha_i$ in $\mathbb{R}^n$ is absolutely convergent in $\mathbb{R}^n$. Let $\alpha_i = \{\alpha_{ki}\}$ and let the norm in $\mathbb{R}^n$, as usual, be given by $\|\beta\| = \left(\sum_{k \leqslant n} |\beta_k|^2 \right)^{\frac{1}{2}}$, $\beta \in \mathbb{R}^n$. If $1 \leqslant k \leqslant n$ we define the increasing sequences σ_k^+ and σ_k^- such that $i \in \sigma_k^{\pm}$ when ever $\alpha_{ki} \gtrless 0$ respectively. Since $\sum_{i=1}^{\infty} \alpha_i$ is subseries convergent there are constants $M_k^{\pm} > 0$ for which $\sup\limits_{m} \left\| \sum\limits_{i \in \sigma_m \cap \sigma_k^{\pm}} \alpha_i \right\| \leqslant M_k^{\pm}$, where σ_m is the set of integers $\leqslant m$. Then

$$\sum_{i \leqslant m} \|\alpha_i\| \leqslant \sqrt{n} \sum_{k=1}^{n} \sum_{i \leqslant m} |\alpha_{ki}|$$

$$\leqslant 2\sqrt{n} \sum_{k=1}^{n} \max_{\pm} \sum_{i \in \sigma_m \cap \sigma_k^{\pm}} |a_{ki}|$$

$$= 2\sqrt{n} \sum_{k=1}^{n} \max_{\pm} \left| \sum_{i \in \sigma_m \cap \sigma_k^{\pm}} \alpha_{ki} \right|$$

$$\leqslant 2\sqrt{n} \sum_{k=1}^{n} \max_{\pm} \left\| \sum_{i \in \sigma_m \cap \sigma_k^{\pm}} \alpha_i \right\|$$

$$\leqslant 2\sqrt{n} \sum_{k=1}^{n} \max_{\pm} M_k^{\pm} < \infty,$$

and the proof is complete.

In order to prove the Dvoretzky-Rogers theorem which states that in every infinite dimensional Banach space there exists an unconditionally convergent series which is not absolutely convergent, we prove a geometrical lemma about symmetric convex bodies in the Euclidean space $\mathbb{R}^n$ (by a *body* we mean the closure of a bounded open set in $\mathbb{R}^n$). For the rest of this chapter we assume X to be infinite dimensional.

Lemma 3. *Let V be a symmetric convex body in $\mathbb{R}^n$. Then there are points $\beta_1, ..., \beta_n$ on the boundary of V such that $\lambda_r^{-1} \sum\limits_{i \leqslant r} \alpha_i \beta_i \in V$ for any $r \leqslant n$ and any $\alpha \neq 0$ in $\mathbb{R}^n$, where $\lambda_r = \|\alpha\| [1 + (r(r-1)/n)^{\frac{1}{2}}]$.*

Proof. Let $u_1, ..., u_n$ be an orthonormal basis for $\mathbb{R}^n$ and let $\alpha = \sum\limits_{i \leqslant n} \alpha_i u_i, \alpha \in \mathbb{R}^n$. We inscribe in V the ellipsoid E of maximum volume, i.e. the set $E = \{\alpha | \alpha \in \mathbb{R}^n, (A\alpha, \alpha) \leqslant 1\}$ for which $|\det A|^{-1}$ is maximal, where A is a positive definite symmetric linear operator of $\mathbb{R}^n$. There is an orthogonal transformation U of $\mathbb{R}^n$ such that the corresponding matrix of $U A U^{-1}$ is diagonal with diagonal elements $a_1, ..., a_n > 0$. Thus $U(E) = \{\alpha | \alpha \in \mathbb{R}^n, \sum\limits_{i \leqslant n} a_i \alpha_i^2 \leqslant 1\}$ is an ellipsoid of maximum volume in $U(V)$. Next, the linear transformation T of $\mathbb{R}^n$ whose corresponding matrix is $\{a_i^{\frac{1}{2}} \delta_{ij}\}$, maps $U(E)$ onto the unit ball B of $\mathbb{R}^n$ and $U(V)$ onto $TU(V)$ which is again a symmetric convex body, and B is an ellipsoid of maximum volume in $TU(V)$. Since $\lambda_r^{-1} \sum\limits_{i \leqslant r} \alpha_i \beta_i \in V$ if and only if $\lambda_r^{-1} \sum\limits_{i \leqslant r} \alpha_i TU \beta_i \in TU(V)$, it is clearly sufficient to solve the problem with V replaced by $TU(V)$.

3*

We now suppose that for some $i \leqslant n$ there is an orthonormal basis $u_1, \ldots, u_n$ for $\mathbb{R}^n$ and i points $\beta_1, \ldots, \beta_i$ of contact of $TU(V)$ with B such that with $\beta_k = \sum_{j \leqslant n} \beta_{kj} u_j$, $k = 1, \ldots, i$,

(i) $$\beta_{kk} \geqslant 0, \quad \beta_{kj} = 0, \quad j > k,$$

(ii) $$\sum_{j < k} \beta_{kj}^2 = 1 - \beta_{kk}^2 \leqslant (k-1)/n.$$

Since E has maximum volume in V there is at least one point of contact, say β_1 of B with $TU(V)$. Let first $u_1 = \beta_1$ and let $u_2, \ldots, u_n$ be any vectors completing an orthonormal basis in $\mathbb{R}^n$. Thus the above conditions are satisfied for $i = 1$. To show it for all $i \leqslant n$ we assume it for i and prove it for $i + 1$.

For $\varepsilon > 0$ we consider the ellipsoid E_ε of points in $\mathbb{R}^n$ which satisfy

$$(1+\varepsilon)^{n-i} \sum_{j \leqslant i} \alpha_j^2 + (1+\varepsilon+\varepsilon^2)^{-i} \sum_{j=i+1}^{n} \alpha_j^2 \leqslant 1. \tag{1}$$

Since the "volume" ratio of E_ε to B is $[(1+\varepsilon+\varepsilon^2)/(1+\varepsilon)]^{i(n-i)} > 1$, we infer that there exists a point α_ε on the boundary of $TU(V)$ which is in E_ε. Because B is contained in $TU(V)$, α_ε satisfies $\sum_{j \leqslant n} \alpha_{\varepsilon j}^2 \geqslant 1$. Together with (1) this gives

$$[(1+\varepsilon)^{n-i} - 1] \sum_{j \leqslant i} \alpha_{\varepsilon j}^2 + [(1+\varepsilon+\varepsilon^2)^{-i} - 1] \sum_{j=i+1}^{n} \alpha_{\varepsilon j}^2 \leqslant 0. \tag{2}$$

Due to the compactness of $TU(V)$, there is a sequence of ε's tending to zero such that α_ε converges to some point β_{i+1} of contact of the boundaries of B and $TU(V)$. Dividing (2) by ε we have in the limit

$$(n-i) \sum_{j \leqslant i} \beta_{i+1,j}^2 - i \sum_{j=i+1}^{n} \beta_{i+1,j}^2 \leqslant 0. \tag{3}$$

The basic vectors $u_{i+1}, \ldots, u_n$, so far have no influence on the conditions (i) and (ii). So we can choose u_{i+1} orthogonal to $u_1, \ldots, u_i$ in $\mathrm{sp}\{u_1, \ldots, u_i, \beta_{i+1}\}$ and we may complete the set $\{u_1, \ldots, u_{i+1}\}$ to a new orthonormal basis of $\mathbb{R}^n$. It is clear that the coordinates of β_j, $j \leqslant i$ do not change and that, under this coordinate transformation, the values of the two sums in (3), as well as that of $\sum_{j \leqslant n} \beta_{i+1,j}^2$ are invariant. Using this, the evident equations $\beta_{i+1,j} = 0$ for $j > i+1$, and the equation $\sum_{j \leqslant i} \beta_{i+1,j}^2 = 1$, (i) and (ii) follow immediately for i replaced by $i+1$, hence for all $i \leqslant n$ (from the fact that the problem is symmetric with respect to the origin it is clear that β_i may be chosen such that β_{ii} is non-negative for all $i \leqslant n$).

Having shown the existence of points $\beta_1, \ldots, \beta_n$ on the boundary of $TU(V)$, satisfying (i) and (ii), let α be a vector of $\mathbb{R}^r$. Then, using (i), (ii) and Schwarz's inequality, one obtains

$$
\begin{aligned}
\left\| \sum_{i \leqslant r} \alpha_i (\beta_i - U_i) \right\| &\leqslant \sum_{i \leqslant r} |\alpha_i| \, \|\beta_i - u_i\| \\
&= \sum_{i \leqslant r} |\alpha_i| \left[\sum_{j < i} \beta_{ij}^2 + (1 - \beta_{ii})^2 \right]^{\frac{1}{2}} \\
&\leqslant \sum_{i \leqslant r} |\alpha_i| \left[\sum_{j < i} \beta_{ij}^2 + 1 - \beta_{ii}^2 \right]^{\frac{1}{2}} \\
&\leqslant \sum_{i \leqslant r} |\alpha_i| \, [2(i-1)/n]^{\frac{1}{2}} \\
&\leqslant \left[\sum_{i \leqslant r} |\alpha_i|^2 \sum_{j \leqslant r} 2(j-1)/n \right]^{\frac{1}{2}} = \left[\frac{r(r-1)}{n} \right]^{\frac{1}{2}} \|\alpha\|.
\end{aligned}
$$

Hence with $\lambda_r = \|\alpha\| \left[1 + (r(r-1)/n)^{\frac{1}{2}} \right]$ we get

$$
\begin{aligned}
\left\| \sum_{i \leqslant r} \alpha_i \beta_i \right\| &\leqslant \left\| \sum_{i \leqslant r} \alpha_i u_i \right\| + \left\| \sum_{i \leqslant r} \alpha_i (\beta_i - u_i) \right\| \\
&\leqslant \left[\sum_{i \leqslant r} |\alpha_i|^2 \right]^{\frac{1}{2}} + \left[\frac{r(r-1)}{n} \right]^{\frac{1}{2}} \|\alpha\| = \lambda_r
\end{aligned}
$$

and the lemma is proved.

Lemma 4. *Let* $c_1, \ldots, c_r$ *be any positive numbers. Then there exists points* $x_1, \ldots, x_r$ *in* X *with* $\|x_i\|^2 = c_i$ *for* $i = 1, \ldots, r$ *and such that*

$$
\left\| \sum_{i \in \mu_r} x_i \right\|^2 \leqslant 4 \sum_{i \in \mu_r} c_i,
$$

where μ_r *is any subset of the numbers* $1, \ldots, r$.

Proof. We take $n = r(r-1)$ and choose a set of n linearly independent elements $y_1, \ldots, y_n$ of X. Then $V = \left\{ \alpha \mid \alpha \in \mathbb{R}^n, \left\| \sum_{j \leqslant n} \alpha_j y_j \right\| \leqslant 1 \right\}$ is a symmetric convex body in $\mathbb{R}^n$, because for $\alpha^+, \alpha^- \in V$ and t in $[0, 1]$, $\left\| \sum_{j \leqslant n} t \alpha_j^+ y_j + \sum_{j \leqslant n} (1-t) \alpha_j^- y_j \right\| \leqslant [t + (1-t)] \max_{\pm} \left\| \sum_{j \leqslant n} \alpha_j^{\pm} y_j \right\| \leqslant 1$ (the symmetry is evident). Let $\beta_1, \ldots, \beta_r$ be r points on the boundary of V such that $(2\|\alpha\|)^{-1} \sum_{j \leqslant r} \alpha_j \beta_j \in V$ for any α in $\mathbb{R}^r$, where the existence of the set $\{\beta_j\}$ is given by the preceding lemma. Let $\lambda_{\mu_r} = 2 \left(\sum_{i \in \mu_r} c_i \right)^{\frac{1}{2}}$. Defining

$$
x_i = c_i^{\frac{1}{2}} \sum_{j \leqslant n} \beta_{ij} y_j, \qquad i \leqslant r,
$$

it is clear that $\|x_i\|^2 = c_i$. Since $\lambda_{\mu_r}^{-1} \sum_{i \in \mu_r} c_i^{\frac{1}{2}} \beta_i$ is in V, we finally get

$$\left\| \sum_{i \in \mu_r} x_i \right\|^2 = \left\| \sum_{i \in \mu_r} c_i^{\frac{1}{2}} \sum_{j \leq n} \beta_{ij} y_j \right\|^2 = \lambda_{\mu_r}^2 \left\| \sum_{j \leq n} \left[\lambda_{\mu_r}^{-1} \sum_{i \in \mu_r} c_i^{\frac{1}{2}} \beta_i \right] y_j \right\|^2 \leq \lambda_{\mu_r}^2 = 4 \sum_{i \in \mu_r} c_i.$$

Remark. Before we prove the next theorem we remember that there exist sequences $\{c_i\}$ of positive terms such that $\sum_{i=1}^{\infty} c_i < \infty$, $\sum_{i=1}^{\infty} c_i^{\frac{1}{2}} = \infty$ and such that $\sum_{i=1}^{\infty} \left(\sum_{j=n_i+1}^{n_{i+1}} c_j \right)^{\frac{1}{2}} < \infty$ for some strictly increasing sequence $\{n_1, n_2, n_3, \ldots\}$ of integers (used in the proof of the following theorem): Let $c_i = (i)^{-2}$ and let $n_i = 2^{i-1}$, $i \geq 1$. Then $\sum_{i=1}^{\infty} c_i = \pi^2/6$, $\sum_{i=1}^{\infty} c_i^{\frac{1}{2}} = 1 + \frac{1}{2} + \frac{1}{3} + \cdots = \infty$ and $\sum_{i=1}^{\infty} \left(\sum_{j=n_i+1}^{n_{i+1}} c_j \right)^{\frac{1}{2}} \leq \sum_{i=1}^{\infty} (2^{i-1} 2^{-2(i-1)})^{\frac{1}{2}} = (1 - 1/\sqrt{2})^{-1} < \infty.$

Theorem 5. *Let $\{c_i\}$ and $\{n_i\}$ be one of the objects defined above. Then there exists an unconditionally convergent series $\sum_{i=1}^{\infty} x_i$ in X with $\|x_i\|^2 = c_i$ for $i = 1, 2, \ldots$.*

Proof. By the preceding lemma we can choose a sequence $\{x_i\}$ in X such that $\|x_i\|^2 = c_i$ and $\left\| \sum_{i \in \sigma_k} x_i \right\|^2 \leq 4 \sum_{i \in \sigma_k} c_i$, $k = 1, 2, \ldots$, where σ_k' is any subsequence of $\sigma_k = \{n_k + 1, \ldots, n_{k+1}\}$.

Let σ be any increasing sequence of integers. Choosing v so large that for $\varepsilon > 0$,

$$\sum_{k=v}^{\infty} \left(\sum_{i \in \sigma_k} c_i \right)^{\frac{1}{2}} < \varepsilon/2,$$

we have for any $\sigma_{pq} = \{p, p+1, \ldots, q\}$ with $q \geq p > n_v$,

$$\left\| \sum_{i \in \sigma_{pq} \cap \sigma} x_i \right\| = \left\| \sum_{k=v}^{\infty} \sum_{i \in \sigma_{pq} \cap \sigma_k \cap \sigma} x_i \right\|$$

$$\leq \sum_{k=v}^{\infty} \left\| \sum_{i \in \sigma_{pq} \cap \sigma_k \cap \sigma} x_i \right\|$$

$$\leq \sum_{k=v}^{\infty} \left(4 \sum_{i \in \sigma_{pq} \cap \sigma_k \cap \sigma} c_i \right)^{\frac{1}{2}}$$

$$\leq 2 \sum_{k=v}^{\infty} \left(\sum_{i \in \sigma_k} c_i \right)^{\frac{1}{2}} < \varepsilon.$$

Finally, since X is complete, this shows that the series $\sum\limits_{i=1}^{\infty} x_i$ is subseries, and hence unconditionally convergent. This finishes the proof of the theorem.

The following main theorem is now an immediate consequence of Theorem 5.

Theorem 6. (DVORETZKY-ROGERS) *The unconditionally convergent series coincide with the absolutely convergent series in the Banach space X if and only if X is finite dimensional.*

Proof. By Theorem 5 there is an unconditionally convergent series $\sum\limits_{i=1}^{\infty} x_i$ such that $\sum\limits_{i=1}^{\infty} \|x_i\| = \sum\limits_{i=1}^{\infty} c_i^{\frac{1}{2}} = \infty$, hence which is not absolutely convergent. On the other hand, if X is finite dimensional, then by Theorem 2 the unconditionally convergent series coincide with the absolutely convergent series.

References for Chapter II: DAY [2], DVORETZKY and ROGERS [1], HILDEBRANDT [1] and HILLE and PHILLIPS [1].

CHAPTER III

Bases for Banach Spaces

Throughout this chapter (and the next two) the basic space X will be a Banach space. According to the three most common used topologies, there are bases for the strong, the weak and the weak* topologies for X, whose definitions are given in the first paragraph. It is shown that every basis for X is a Schauder basis, a basis with continuous linear coefficient functionals. The next paragraph shows under which conditions a biorthogonal system is a basis for X, the equivalence of strong and weak Schauder bases for X and relations between bases for X and bases for the adjoint space X^*. Three paragraphs are devoted to retro-, shrinking, boundedly complete, unconditional, absolutely convergent and uniform bases. Some applications of summability methods on the theory of bases are given in the sixth section and in the last paragraph bases for the special spaces c_0, $l_p(1 \leqslant p < \infty)$, $C[0,1]$, $L_p[0,1]$ $(1 \leqslant p < \infty)$, $L_2[0,2\pi]$ and A^2 are considered.

1. Bases Corresponding to Different Topologies

In this section we shall be concerned with the definitions and some basic facts pertaining to bases for Banach spaces in the different possible topologies. Let X be a Banach space over the field Φ of real (or complex) numbers and let X^* be its conjugate space. The topologies considered are the strong topology (=norm topology of X), the weak topology (=X^* topology of X) and, when the Banach space is a conjugate space, the weak* topology (=X topology of X^*).

Definition 1. *A (weak, weak*) basis for a Banach space X over the field Φ is a sequence $\{x_i\}$ in X such that to every element x in X there corresponds a unique sequence $\{\alpha_i\}$ in Φ for which $x = \lim\limits_{n} \sum\limits_{i \leqslant n} \alpha_i x_i$ in the strong (weak, weak* respectively) topology.*

The elements in $\{\alpha_i\}$ naturally depend linearly on x and they are called the *coefficient functionals* of the basis $\{x_i\}$. Moreover, that each α_i is a unique function of x implies that every element in the sequence $\{x_i\}$ is non-zero.

Definition 2. *A (weak, weak*) basis which has (weakly, *-weakly) continuous coefficient functionals is said to be a (weak, weak*) Schauder basis.*

It will soon be shown that every weak basis for X is a basis for X and that every basis for X is a Schauder basis for X. However, that every weak* basis for a conjugate space X^* is a weak* Schauder basis for X^* is not the case as will be clear by an example appearing later in section 7 (Theorem 4).

Theorem 3. *In a Banach space X every basis $\{x_i\}$ for X is a Schauder basis for X.*

Proof. Let Y be the vector space over the field Φ of all sequences $y = \{\alpha_i\}$, $\alpha_i \in \Phi$ for which $\lim_n \sum_{i \leqslant n} \alpha_i x_i$ exists. Hence $\sup_n \left\| \sum_{i \leqslant n} \alpha_i x_i \right\| < \infty$ for each $y \in Y$. Clearly, the function $\|y\|$ on Y, given by $\|y\| = \sup_n \left\| \sum_{i \leqslant n} \alpha_i x_i \right\|$ defines a norm on Y and supplied with this norm, Y becomes a normed linear space. Now, let $T: Y \to X$ be the linear transformation defined by $x = \lim_n \sum_{i \leqslant n} \alpha_i x_i$. Since every x in X has a unique expansion of this form, T is one-to-one and onto. Obviously $\|x\| = \|Ty\| \leqslant \|y\|$ so that T is continuous. As we show later Y is complete and thus a Banach space. Hence T is a topological isomorphism (Theorem I.2.6) and we have $|\alpha_n| \, \|x_n\| = \left\| \sum_{i \leqslant n} \alpha_i x_i - \sum_{i \leqslant n-1} \alpha_i x_i \right\| \leqslant 2\|y\| \leqslant 2\|T^{-1}\| \, \|x\|$. From this and from the uniqueness of the expansion coefficients α_i we conclude that each α_n is a continuous linear functional on X.

It remains to show that Y is complete. Let $\{y_p\}$ be a Cauchy sequence in Y (each y defining a sequence $\{\alpha_{pi}\} \subset \Phi$). Since for all i, $|\alpha_{pi} - \alpha_{qi}| \, \|x_i\| \leqslant 2 \sup_n \left\| \sum_{i \leqslant n} (\alpha_{pi} - \alpha_{qi}) x_i \right\| = 2\|y_p - y_q\|$, and since Φ is complete, there is a sequence $\{\alpha_i\} \subset \Phi$ such that $\lim_p \alpha_{pi} = \alpha_i$. Given $\varepsilon > 0$, there is an index r such that $\|y_p - y_r\| < \varepsilon/3$ for all $p \geqslant r$. Hence $\left\| \sum_{i \leqslant n} (\alpha_{pi} - \alpha_{ri}) x_i \right\| < \varepsilon/3$, $p \geqslant r$, uniformly for all n. Taking the limit on p we obtain $\sup_n \left\| \sum_{i \leqslant n} (\alpha_i - \alpha_{ri}) x_i \right\| \leqslant \varepsilon/3$. Now, since $y_r \in Y$, there is an index n_ε', depending on r, such that $\left\| \sum_{i=n}^{m} \alpha_{ri} x_i \right\| < \varepsilon/3$, $m \geqslant n \geqslant n_\varepsilon$. Hence for each $m, n \geqslant n_\varepsilon$ and $m \geqslant n$ we have

$$\left\|\sum_{i=n}^{m}\alpha_i x_i\right\| \leqslant \left\|\sum_{i=n}^{m}(\alpha_i-\alpha_{ri})x_i\right\| + \left\|\sum_{i=n}^{m}\alpha_{ri}x_i\right\| < \varepsilon.$$ Consequently, $y=\{\alpha_i\}$ is an
element of Y and by what has preceded, $y=\lim_p y_p$. Y is complete, there-fore, and the theorem is verified.

Corollary 4. *Every weak basis for X is a weak Schauder basis for X.*

Proof. Y is defined similarly as in the proof of the foregoing theorem, with $\lim_n \sum_{i\leqslant n}\alpha_i x_i = x$ in the weak topology for X. By (I.3.15), again, $\sup_n\left\|\sum_{i\leqslant n}\alpha_i x_i\right\| < \infty$. Since it will be apparent that Y is still a Banach space, the first part of the proof for the theorem applies also in this situation (where T is, naturally, defined by a weak limit). Thus each α_n is again a strongly (and hence a weakly (I.3.3)) continuous linear functional on X.

If $\{y_p\}$ is a Cauchy sequence in Y we have again a sequence $\{\alpha_i\}$ in Φ such that $\lim_p \alpha_{pi}=\alpha_i$. The continuity of T (by (I.3.15), $\|Ty\| \leqslant \sup_n\left\|\sum_{i\leqslant n}\alpha_i x_i\right\| = \|y\|$ with $y=\{\alpha_i\}$) and the completeness of X imply that Ty_p converges, say, to an element x in X. Given $\varepsilon>0$, there is an index r such that $\|y_p-y_q\| < \varepsilon/3$ for all $p,q\geqslant r$. Hence $\sup_n\left\|\sum_{i\leqslant n}(\alpha_{pi}-\alpha_{qi})x_i\right\| < \varepsilon/3$ for $p,q\geqslant r$ and it follows that $\sup_n\left\|\sum_{i\leqslant n}(\alpha_i-\alpha_{qi})x_i\right\| \leqslant \varepsilon/3$, $q\geqslant r$. Now, there is a fixed index $q\geqslant r$ for which $\|x-Ty_q\| < \varepsilon/3$ and for each $x^*\in X^*$ with $\|x^*\|\leqslant 1$ there is an n_ε, depending on x^* and on q, such that $\left|x^*\left(\sum_{i\leqslant n}\alpha_{qi}x_i - Ty_q\right)\right| < \varepsilon/3$, $n\geqslant n_\varepsilon$. Hence for every $n\geqslant n_\varepsilon$ we have
$$\left|x^*\left(\sum_{i\leqslant n}\alpha_i x_i - x\right)\right| \leqslant \left\|\sum_{i\leqslant n}(\alpha_i-\alpha_{qi})x_i\right\| + \left|x^*\left(\sum_{i\leqslant n}\alpha_{qi}x_i - Ty_q\right)\right| + \|Ty_q-x\| < \varepsilon.$$
Thus Y is complete, since $\{\alpha_i\}$ is in Y and is the strong limit of $\{\alpha_{pi}\}$ in Y, and so we are done.

Remark. A generalization of the above theorem to complete metric linear spaces with translation-invariant metric, given in the last chapter (Theorem IX.5.2), deserves mention. That is, the local convexity hypo-thesis in the theorem may be omitted.

2. Biorthogonal Systems

Biorthogonal systems play a central role in the theory of bases, since every basis for a Banach space X, together with its associated sequence of coefficient functionals constitutes a biorthogonal system. On the other hand, provided some conditions are satisfied, a biorthogonal system is a basis for some closed linear subspace of X.

Definition 1. *Let $\{x_i\}$ and $\{x_i^*\}$ be sequences in X and X^* respectively. $\{x_i, x_i^*\}$ is called a biorthogonal system for X if $x_i^*(x_j) = \delta_{ij}$. The endomorphisms U_n of X, defined by $U_n x = \sum_{i \leq n} x_i^*(x) x_i$ for all n and all x in X, are called expansion operators.*

Evidently, the operators U_n are projections of X with the properties $U_m U_n = U_{\min\{m,n\}}$.

The following theorem was first proved by Banach, but based on stronger assumptions as stated here:

Theorem 2. *Let $\{x_i, x_i^*\}$ be a biorthogonal system for X such that $\sup|x^*(U_n x)| < \infty$, $x \in X$, $x^* \in X^*$. Then $\{x_i\}$ is a basis for $\overline{\mathrm{sp}}\{x_i\}$ and $\{x_i^*\}$ is a basis for $\overline{\mathrm{sp}}\{x_i^*\}$.*

Proof. By hypothesis $\sup_n |x^*(U_n x)| < \infty$, $x \in X$, $x^* \in X^*$. Thus there is a constant $M < \infty$ such that $\sup_n \|U_n\| \leq M$ (I.3.14). Let x be in $\overline{\mathrm{sp}}\{x_i\}$. Then we have for every $\varepsilon > 0$ an index j and an element y_j in $\mathrm{sp}\{x_1, \ldots, x_j\}$ such that $\|x - y_j\| < \varepsilon$. But $\|x - U_n x\| \leq \|x - y_j\| + \|y_j - U_n y_j\| + \|U_n y_j - U_n x\|$ and $U_n y_j = y_j$ for each $n \geq j$, in virtue of the biorthogonal relations. Hence $\|x - U_n x\| \leq (1 + M)\varepsilon$ for all $n \geq j$ so that $x = \lim_n U_n x = \lim_n \sum_{i \leq n} x_i^*(x) x_i$. The coefficients $x_i^*(x)$ for the expansion of x are unique, since $\lim_n \sum_{i \leq n} \alpha_i x_i = 0$, $\alpha_i \in \Phi$, $i = 1, 2, \ldots$, as is easy to see by multiplication with x_j^* and by use of $x_j^*(x_i) = \delta_{ij}$, implies $\alpha_j = 0$ for all j.

On the other hand, let x^* be in $\overline{\mathrm{sp}}\{x_i^*\}$. Since $\|U_n^*\| = \|U_n\|$, in just the same way as above, it follows $x^* = \lim_n U_n^* x^* = \lim_n \sum_{i \leq n} x^*(x_i) x_i^*$. Similarly, the $x^*(x_i)$ are unique and the theorem is proved.

Remark. We observe that Theorem 2 has an obvious generalization (cf. Corollary IX.5.6) to barrelled topological vector spaces, in that situation the proof is then based on the Barrel theorem.

Corollary 3. *Let $\{x_i, x_i^*\}$ be a biorthogonal system for X. Then $\{x_i\}$ is a basis for $\overline{\mathrm{sp}}\{x_i\}$ if and only if $1 \leq \sup_n \|U_n\| < \infty$.*

Proof. The sufficiency follows immediately from the preceding theorem. To prove the necessity, let $\{x_i\}$ be a basis for $\overline{\mathrm{sp}}\{x_i\}$. Then, since $x = \lim_n U_n x$, we have $\|x\| \leq \sup_n \|U_n x\| < \infty$ for all x in X, by (I.3.14) implying $1 \leq \sup_n \|U_n\| < \infty$.

Theorem 4. *$\{x_i\}$ is a (weak) basis for X if and only if there is a sequence $\{x_i^*\}$ in X^* such that $\{x_i, x_i^*\}$ is a biorthogonal system for X and $\sum_{i \leq n} x_i^*(x) x_i$ converges strongly (respectively weakly) to x for each x in X.*

Proof. Let $\{x_i, x_i^*\}$ be a biorthogonal system for X such that $\sum\limits_{i \leqslant n} x_i^*(x)x_i$ converges weakly or strongly to x for each x in X. Then it is clear that $\sup\limits_n \left| \sum\limits_{i \leqslant n} x_i^*(x)x^*(x_i) \right| < \infty$, $x \in X$, $x^* \in X^*$ and by (I.3.15), $\overline{sp}\{x_i\} = X$. Therefore, the foregoing theorem applies and $\{x_i\}$ is a basis for X. Conversely, suppose that $\{x_i\}$ is a (weak) basis for X. Then there exists to each x in X a unique sequence $\{\alpha_i\}$ in Φ such that $\sum\limits_{i \leqslant n} \alpha_i x_i$ converges (weakly) to x. The uniqueness implies the existence of linear functionals x_i^* on X such that $x_i^*(x) = \alpha_i$ for every x in X. From Theorem 1.3 and Corollary 1.4 we know that these functionals are continuous, hence elements of X^*. Moreover, the uniqueness of the sequence $\{x_i^*(x)\}$ implies the biorthogonal relations $x_i^*(x_j) = \delta_{ij}$. Thus $\{x_i, x_i^*\}$ is a biorthogonal system for X and the proof of the theorem is complete.

Remark. From now on we conveniently write for a basis both symbols $\{x_i\}$, as well as $\{x_i, x_i^*\}$. It is practical to use the first one if no reference is made to the associated sequence of coefficient functionals $\{x_i^*\}$ and by use of the second one the introduction and definition of $\{x_i^*\}$ becomes unnecessary.

The following corollary is known as *weak basis theorem* given (without proof) already in Banach's monograph [1, p. 238]. In 1959 it has been established by BESSAGA and PELCZYNSKI [5] for the more general case of a Fréchet space.

Corollary 5. $\{x_i\}$ *is a basis for X if and only if it is a weak basis for X.*

Proof. Let $\{x_i\}$ be a weak basis for X. Then, according to the foregoing theorem, there is a sequence $\{x_i^*\}$ in X^*, biorthogonal to $\{x_i\}$, and such that $\sum\limits_{i \leqslant n} x_i^*(x)x_i$ converges weakly to x for each x in X. As we have shown in the first part of the proof of the theorem, $\{x_i\}$ then is a basis for X. Conversely, since strong convergence implies weak convergence, it follows that a strong basis for X is a weak basis for X and the proof of the corollary is finished.

Theorem 6. *Let $\{x_i, x_i^*\}$ be a biorthogonal system for X. If $\sum\limits_{i \leqslant n} x^*(x_i)x_i^*$ converges to x^* in the weak* topology of X^* for every x^* in X^*, then $\{x_i\}$ constitutes a basis for X.*

Proof. By hypothesis, $\lim\limits_n \sum\limits_{i \leqslant n} x^*(x_i)x_i^*(x) = x^*(x)$ for each x^* in X^* and x in X. This implies $\sup\limits_n \left| \sum\limits_{i \leqslant n} x^*(x_i)x_i^*(x) \right| < \infty$ for each x^* in X^* and x in X so that, by Theorem 2, $\{x_i, x_i^*\}$ is a basis for $\overline{sp}\{x_i\}$. To prove the theorem we only must show that $\overline{sp}\{x_i\} = X$. If we suppose the contrary, then there is an x in X which is not in $\overline{\mathrm{sp}}\{x_i\}$. In this case there exists an x^* in X^* such that $x^*(x)=1$ and $x^*(x_i)=0$ for all i (I.3.6). Consequently, $0=\lim\limits_n \sum\limits_{i\leqslant n} x^*(x_i)x_i^*(x)=x^*(x)=1$ which is the desired contradiction.

Theorem 7. *If $\{x_i,x_i^*\}$ is a basis for X, then $\{x_i^*,Jx_i\}$ is a weak* Schauder basis for X^*. Conversely, if $\{x_i^*,x_i^{**}\}$ is a weak* Schauder basis for X^*, then $\{x_i^{**}\}$ is a sequence in $J(X)$ and $\{J^{-1}x_i^{**},x_i^*\}$ is a basis for X.*

Proof. We recall that J is the natural embedding of X into its second adjoint X^{**}. Let $\{x_i,x_i^*\}$ be a basis for X. Since $Jx(x_i^*)=x_i^*(x_j)=\delta_{ij}$, $\{x_i^*,Jx_i\}$ is a biorthogonal system for X^* and by (I.3.20) each Jx_i is weak* continuous. For each x^* in X^* and x in X it follows that $\lim\limits_n \left[x^* - \sum\limits_{i\leqslant n} Jx_i(x^*)x_i^*\right](x)=\lim\limits_n x^*\left[x - \sum\limits_{i\leqslant n} x_i^*(x)x_i\right]=0$. To prove the uniqueness of the coefficients $Jx_i(x^*)$, we take a sequence $\{\alpha_i\}$ in Φ and assume that $\lim\limits_n \sum\limits_{i\leqslant n} \alpha_i x_i^*(x)=0$ for all x in X. Then with $x=x_j$, we obtain $\alpha_j=0$ for all j and the first part of the theorem is proved.

On the other hand, let $\{x_i^*\}$ be a weak* Schauder basis for X^* with coefficient functionals $x_1^{**},x_2^{**},\ldots$. Then we have $x^*(x) = \lim\limits_n \sum\limits_{i\leqslant n} x_i^{**}(x^*)x_i^*(x)$ for every x^* in X^* and every x in X, and the linear functionals x_i^{**} on X^* are continuous in the weak* topology of X^*. This implies (I.3.20) the existence of elements x_i, $i=1,2,\ldots$, in X such that $x_i^{**}(x^*)=x^*(x_i)$ for every x^* in X^*, and hence that $x_i^{**}\in J(X)$. Consequently, $x^*(x)=\lim\limits_n \sum\limits_{i\leqslant n} x^*(x_i)x_i^*(x)$ for each x^* in X^* and x in X. In other words, $\sum\limits_{i\leqslant n} x_i^*(x)x_i$ converges weakly to x for all x in X. Since by the equation $x_i^*(x_j)=x_j^{**}(x_i^*)$ and by uniqueness of the coefficient sequence $\{x_j^{**}(x_i^*)\}$ belonging to x_i^*, it is clear that $x_i^*(x_j)=\delta_{ij}$, it follows at once from Theorem 4 that $\{x_i,x_i^*\}$ constitutes a weak basis for X, and hence (by Corollary 5) a basis for X.

Corollary 8. *Let X be a non-separable Banach space. Then X^* has no weak* Schauder basis (and no basis either).*

Corollary 9. *Let X be reflexive. If $\{x_i,x_i^*\}$ is a basis for X, then $\{x_i^*,Jx_i\}$ is a basis for X^*. Conversely, if $\{x_i^*,x_i^{**}\}$ is a basis for X^*, then $\{x_i^{**}\}$ is a sequence in $J(X)$ and $\{J^{-1}x_i^{**},x_i^*\}$ is a basis for X.*

Proof. The first part becomes clear from the fact that for reflexive X, the weak* and weak topologies for X^* are equivalent. This implies that $\{x_i^*,Jx_i\}$ is a weak basis for X^* and by Corollary 5, a basis for X^*. Conversely, if $\{x_i^*,x_i^{**}\}$ is a basis for X^*, then it is a weak* Schauder basis for X^* (by (I.3.20) and since strong convergence in X^* implies weak* convergence in X^*), and so $\{J^{-1}x_i^{**}, x_i^*\}$ is a basis for X.

Theorem 10. (WILANSKY) *Let* $\{x_i, x_i^*\}$ *be a basis for* X. *Then* $\sup_n \left\| \sum_{i \leqslant n} x^*(x_i) x_i^* \right\| < \infty$ *and* $\sup_n \left\| \sum_{i \leqslant n} x^{**}(x_i^*) x_i \right\| < \infty$ *for each* x^* *in* X^* *and each* x^{**} *in* X^{**}.

Proof. Corollary 3 shows that $\sup_n \left| \sum_{i \leqslant n} x^*(x_i) x_i^*(x) \right| < \infty$ for all x in X^* and all x in X. As a consequence of the principle of uniform boundedness (I.3.14), we then have $\sup_n \left\| \sum_{i \leqslant n} x^*(x_i) x_i^* \right\| < \infty$. On the other hand,

$$\sup_n \left| \sum_{i \leqslant n} x^{**}(x_i^*) x^*(x_i) \right| \leqslant \sup_n \|x^{**}\| \left\| \sum_{i \leqslant n} x^*(x_i) x_i^* \right\| < \infty \quad \text{for all } x^{**} \text{ in } X^{**}$$

and all x^* in X^*. Finally, the proof of the theorem follows by application of the same principle once more.

Definition 11. *A basis* $\{x_i^*\}$ *for* X^* *is called a retro-basis if its biorthogonal sequence* $\{x_i^{**}\}$ *is contained in* $J(X)$.

It is apparent that each retro-basis for X^* is a weak* Schauder basis for X^*.

Theorem 12. *Let* $\{x_i^*\}$ *be a retro-basis for* X^* *with corresponding biorthogonal sequence* $\{Jx_i\}$. *Then* $\{x_i, x_i^*\}$ *is a basis for* X, *and* $\sum_{i \leqslant n} x^{**}(x_i^*) x_i$ *converges in* X *if and only if* x^{**} *is an element of* $J(X)$.

Proof. By hypothesis, $\{x_i, x_i^*\}$ is á biorthogonal system so that since strong convergence in X^* implies weak* convergence in X^*, Theorem 6 applies and $\{x_i\}$ is a basis for X. Then it is clear that the series given in the theorem converges if x^{**} is in $J(X)$. On the other hand, let the series converge (to an element y in X) and define $y_n = \sum_{i \leqslant n} x^{**}(x_i^*) x_i$. For every x^* in

$$X^*, \; Jy(x^*) = \lim_n J y_n(x^*) = \lim_n \sum_{i \leqslant n} x^{**}(x_i^*) x^*(x_i) = x^{**}\left(\lim_n \sum_{i \leqslant n} J x_i(x^*) x_i^* \right)$$

$= x^{**}(x^*)$ and we finally see that x^{**} is in $J(X)$.

3. Shrinking and Boundedly Complete Bases

Definition 1. *A basis* $\{x_i\}$ *for* X *with associated sequence of expansion operators* $\{U_i\}$ *is shrinking if and only if*
$\lim_n \sup \{ |x^*(x - U_n x)| \mid \|x\| \leqslant 1 \} = 0$ *for each* x^* *in* X^*.

We note that by Corollary 2.3, $\lim_n \sup \{ |x^*(y)| \mid y \in \overline{\text{sp}} \{x_n, x_{n+1}, x_{n+2}, \ldots\}, \|y\| \leqslant 1 \} = 0$ for each x^* in X^* is an equivalent condition for shrinking. Furthermore, an equivalent name for a shrinking basis $\{x_i\}$ is that of a *weakly uniform basis*, since it will be immediately clear from the definition that the expansion $U_n x$ for x converges in the weak topology of X, uniformly on the unit ball of X.

Definition 2. *A basis* $\{x_i\}$ *for* X *is said to be monotone if* $\|U_n x\|$ *is a non-decreasing function of* n *for all* x *in* X.

Theorem 3. *A basis* $\{x_i\}$ *for* X *is monotone if and only if* $\sup_n \|U_n\| \leqslant 1$.

Proof. We assume $\{x_i\}$ to be not monotone. Then there is an n and an x such that $\|U_{n+1} x\| < \|U_n x\|$. But this implies $\|U_n\| \geqslant \|U_n U_{n+1} x\| / \|U_{n+1} x\| = \|U_n x\| / \|U_{n+1} x\| > 1$. This shows that a basis $\{x_i\}$ with $\sup_n \|U_n\| \leqslant 1$ is monotone. The converse is obvious.

Theorem 4. *Let* $\{x_i, x_i^*\}$ *be a basis for* X. *Then the following two statements are equivalent:*

(i) $\{x_i\}$ *is a shrinking basis for* X.
(ii) $\{x_i^*\}$ *is a basis for* X^*.

Moreover, $\{x_i\}$ *is monotone if and only if* $\{x_i^*\}$ *is also monotone.*

Proof. (i)$\Rightarrow$(ii). If $\{x_i, x_i^*\}$ is shrinking, then for all $x^* \in X^*$,
$$0 = \lim_n \sup \{|x^*(x - U_n x)| \mid \|x\| \leqslant 1\} = \lim_n \sup \{|(I - U_n^*) x^*(x)| \mid \|x\| \leqslant 1\}$$
$$= \lim_n \|x^* - U_n^* x^*\| = \lim_n \left\| x^* - \sum_{i \leqslant n} x^*(x_i) x_i^* \right\| = \lim_n \left\| x^* - \sum_{i \leqslant n} J x_i(x^*) x_i^* \right\|.$$
Since $J x_i(x_j^*) = x_j^*(x_i) = \delta_{ij}$, it follows from Theorem 2.4 that $\{x_i^*, J x_i\}$ is a basis for X^*.

(ii)$\Rightarrow$(i). Suppose that $\{x_i^*, x_i^{**}\}$ is a basis for X^*. Then
$$x^*(x_i) = \lim_n \sum_{j \leqslant n} x_j^{**}(x^*) x_j^*(x_i) = x_i^{**}(x^*)$$

for every $x^* \in X^*$ and for all i. Hence $x_i^{**} = J x_i$ and so
$$0 = \lim_n \left\| x^* - \sum_{i \leqslant n} J x_i(x^*) x_i^* \right\| = \lim_n \sup \left\{ \left| x^*(x) - \sum_{i \leqslant n} x^*(x_i) x_i^*(x) \right| \mid \|x\| \leqslant 1 \right\}$$
$$= \lim_n \sup \{|x^*(x - U_n x)| \mid \|x\| \leqslant 1\}$$

which shows that the basis $\{x_i\}$ is shrinking. Since
$$\sup_n \left\{ \left\| \sum_{i \leqslant n} x_i^{**}(x^*) x_i^* \right\| \mid \|x^*\| \leqslant 1 \right\}$$
$$= \sup_n \sup \left\{ \left| \sum_{i \leqslant n} J x_i(x^*) x_i^*(x) \right| \mid \|x\| \leqslant 1, \|x^*\| \leqslant 1 \right\}$$
$$= \sup_n \sup \left\{ \left| x^* \left(\sum_{i \leqslant n} x_i^*(x) x_i \right) \right| \mid \|x\| \leqslant 1, \|x^*\| \leqslant 1 \right\}$$
$$= \sup_n \sup \left\{ \left\| \sum_{i \leqslant n} x_i^*(x) x_i \right\| \mid \|x\| \leqslant 1 \right\} = \sup_n |U_n\|,$$

the proof is complete.

Corollary 5. *Let* $\{x_i, x_i^*\}$ *be a basis for* X. *If this basis is shrinking, then* $\{x_i^*, J x_i\}$ *is a basis for* X^*. *Conversely, if* $\{x_i^*, x_i^{**}\}$ *is a basis for* X^*, *then* $x_i^{**} = J x_i$ *for all* i.

Corollary 6. *A basis* $\{x_i, x_i^*\}$ *for* X *is shrinking if and only if* $\overline{\mathrm{sp}}\,\{x_i^*\} = X^*$.

Proof. The necessity is clear from Corollary 5. Conversely, let $\overline{\mathrm{sp}}\,\{x_i^*\} = X^*$. From Corollary 2.3 we infer that $\sup_n |x^*(U_n x)| < \infty$, $x \in X, x^* \in X^*$ so that by Theorem 2.2, $\{x_i^*\}$ is a basis for X^*. This finally shows that the basis $\{x_i\}$ is shrinking.

Theorem 7. *A basis* $\{x_i, x_i^*\}$ *for* X *is shrinking if and only if a bounded sequence* $\{y_k\}$ *in* X *converges weakly to zero whenever* $\lim_k x_i^*(y_k) = 0$ *for* $i = 1, 2, \dots$.

Proof. Without loss of generality we can take $\{y_k\}$ in X such that $\sup_k \|y_k\| \leqslant 1$. Let us first assume that the basis is shrinking. Then for each $x^* \neq 0$ in X^* there is for every $\varepsilon > 0$ an index n such that $\sup \{|x^*(x - U_n x)| \mid \|x\| \leqslant 1\} < \varepsilon/2$. Also, by hypothesis, there is an index k_0 such that for $k \geqslant k_0, |x_i^*(y_k)| < \varepsilon \left(2\|x^*\| \sum_{j \leqslant n} \|x_j\|\right)^{-1}, i = 1, \dots, n$. Hence

$$|x^*(y_k)| \leqslant |x^*(y_k - U_n y_k)| + |x^*(U_n y_k)|$$
$$< \varepsilon/2 + \left|x^*\left(\sum_{i \leqslant n} x_i^*(y_k) x_i\right)\right|$$
$$\leqslant \varepsilon/2 + \|x^*\| \sum_{i \leqslant n} |x_i^*(y_k)| \, \|x_i^*\| < \varepsilon$$

or, $\lim_k x^*(y_k) = 0$ for every x^* in X^*.

To show the sufficiency, we suppose that $\{x_i, x_i^*\}$ is not shrinking. Then there exists an $\varepsilon > 0$, an x^* in X^*, a sequence $\{y_k\}$ in X and a sequence $\{m_k\}$ of strictly increasing integers such that $x^*(y_k) > \varepsilon$ and $y_k \in \overline{\mathrm{sp}}\,\{x_{m_k+1}, \dots, x_{m_{k+1}}\}$. Thus $\lim_k x_i^*(y_k) = 0$ for $i = 1, 2, \dots$. But y_k converges not weakly to zero so that the basis must be shrinking.

Definition 8. *A basis* $\{x_i\}$ *for* X *is boundedly complete if for each sequence* $\{\alpha_i\}$ *in* Φ *for which* $\sup_n \left\|\sum_{i \leqslant n} \alpha_i x_i\right\| < \infty$, *there exists an* x *in* X *such that* $x = \lim_n \sum_{i \leqslant n} \alpha_i x_i$.

Theorem 9. *Let* $\{x_i, x_i^*\}$ *be a shrinking basis for* X. *Then* $\{x_i^*\}$ *is a boundedly complete basis for* X^*.

Proof. Let $\{\alpha_i\} \subset \Phi$ be a sequence such that $\sup_n \|y_n^*\| < \infty$, where $y_n^* = \sum_{i \leqslant n} \alpha_i x_i^*$. According to Corollary 5, $\{x_i^*, J x_i\}$ is a basis for X^* and it is then clear that $\lim_{n} y_n^*(x_i) = \alpha_i$. Thus the Banach-Steinhaus theorem applies so that there exists a y^* in X^* with $\|y^*\| \leqslant \sup_{n} \|y_n^*\| < \infty$, such that $\lim_{n} y_n^*(x) = y^*(x)$ for all x in X. Consequently, $y^* = \lim_{m} \sum_{j \leqslant m} y^*(x_j)x_j^*$

$= \lim_{m} \sum_{j \leqslant m} x_j^* \lim_{n} y_n^*(x_j) = \lim_{m} \sum_{i \leqslant m} \alpha_i x_i^*$ in the strong topology for X^*.

This concludes the proof of the theorem since we have shown that the basis $\{x_i^*\}$ is boundedly complete.

Theorem 10. *If $\{x_i, x_i^*\}$ is a boundedly complete basis for X, then $\sum_{i \leqslant n} x^{**}(x_i^*)x_i$ converges in X for each x^{**} in X^{**}.*

Proof. For each y^{**} in X^{**} there is a sequence $\{y_i\}$ in X for which $y^{**}(x^*) = \lim_{j} J y_j(x^*) = \lim_{j} x^*(y_j)$, $x^* \in X^*$, because by I.3.22, $J(X)$ is dense in X^{**} in the weak* topology of X^{**}. Moreover, from the weak convergence of y_j we obtain (I.3.15) that $\sup_{j} \|y_j\| < \infty$. Then

$\sup_{n} \left\| \sum_{i \leqslant n} y^{**}(x_i^*)x_i \right\| = \sup_{n} \left\| \lim_{j} \sum_{i \leqslant n} x_i^*(y_j)x_i \right\| \leqslant \sup_{n} \|U_n\| \cdot \sup_{j} \|y_j\| < \infty$ for every y^{**} in X^{**}, where U_n is defined in 2.1 and where we have used Corollary 2.3. This, and the bounded completeness of the basis for X imply the convergence of $\sum_{i \leqslant n} y^{**}(x_i^*)x_i$ in X.

Theorem 11. *Let $\{x_i, x_i^*\}$ be a basis for X such that $\sum_{i \leqslant n} x^{**}(x_i^*)x_i$ converges weakly to an element of X for each x^{**} in X^{**}. Then X is topologically isomorphic to a conjugate space.*

Proof. We define $\overline{\text{sp}} \{x_i^*\}^{\perp} = \{x^{**} | x^{**} \in X^{**}, x^{**}(x^*) = 0$ for all x^* in $\overline{\text{sp}} \{x_i^*\}\}$. Let x^{**} be in $\overline{\text{sp}} \{x_i^*\}^{\perp} \cap J(X)$. Then for every x^* in X^*, $x^*(J^{-1}x^{**}) = \lim_{n} \sum_{i \leqslant n} x_i^*(J^{-1}x^{**})x^*(x_i) = \lim_{n} \sum_{i \leqslant n} x^{**}(x_i^*)x^*(x_i) = 0$ so that $x^{**} = 0$, or $\overline{\text{sp}} \{x_i^*\}^{\perp} \cap J(X) = \{0\}$. Moreover, for each x^{**} in X^{**} there is an x in X such that $\lim_{n} \sum_{i \leqslant n} x^{**}(x_i^*)x^*(x_i) = x^*(x)$, $x^* \in X^*$. We have $[x^{**} - Jx](x^*) = x^{**}(x^*) - x^*(x) = x^{**}(x^*) - \lim_{n} \sum_{i \leqslant n} x^{**}(x_i^*)x^*(x_i) = 0$ for each $x^* = x_i^*$, $i = 1, 2, \ldots$ and hence for each x^* in $\overline{\text{sp}} \{x_i^*\}$. Therefore, $X^{**} = J(X) \oplus \overline{\text{sp}} \{x_i^*\}^{\perp}$. Consequently, there is a projection of X^{**} on $J(X)$ and (I.2.13) $J(X)$ is topologically isomorphic to the factor space $X^{**}/\overline{\text{sp}} \{x_i^*\}^{\perp}$. By I.3.8 it finally follows that the last quantity is topologically isomorphic to $\overline{\text{sp}} \{x_i^*\}^*$. This concludes the proof, since then X is topologically isomorphic to $\overline{\text{sp}} \{x_i^*\}^*$.

Corollary 12. *If X has a boundedly complete basis, then X is topologically isomorphic to a conjugate space.*

4. Unconditional Bases

Definition 1. *A basis* $\{x_i, x_i^*\}$ *for* X *is said to be unconditional if for all* x *in* X *the series* $\sum_{i=1}^{\infty} x_i^*(x)x_i$ *converges unconditionally.*

Now, let Σ be the set of all finite subsets μ of the set of all positive integers and let $\{x_i\}$ be an unconditional basis for X. Furthermore, let Y be the vector space of all sequences $y=\{\alpha_i\}, \alpha_i \in \Phi$, for which $\sum_{i \leqslant n} \alpha_i x_i$ converges to an element x in X. Let S be the set of all sequences $\{\gamma_i\}$ in Φ with $|\gamma_i| \leqslant 1$. Since for each x^* in X^* the sum $\sum_{i=1}^{\infty} \alpha_i x^*(x_i)$ is absolutely convergent (Theorem II.2.2), $\sum_{i=1}^{\infty} |\alpha_i x^*(x_i)|$ is finite and an upper bound for $\sup\left\{\left|\sum_{i \in \mu} \gamma_i \alpha_i x^*(x_i)\right| \,\middle|\, \{\gamma_i\} \in S, \mu \in \Sigma\right\}$. The principle of uniform boundedness (I.3.14) then implies that $\sup\left\{\left\|\sum_{i \in \mu} \gamma_i \alpha_i x_i\right\| \,\middle|\, \{\gamma_i\} \in S, \mu \in \Sigma\right\} < \infty$. Next, we define a norm in Y by $\|y\| = \sup\left\{\left\|\sum_{i \in \mu} \gamma_i \alpha_i x_i\right\| \,\middle|\, \{\gamma_i\} \in S, \mu \in \Sigma\right\}$, and show that Y is complete:

Lemma 2. *Y is a Banach space and is topologically isomorphic with X.*

Proof. Let $T: Y \to X$ be given by $Ty = \lim_n \sum_{i \leqslant n} \alpha_i x_i$, $y \in Y$. Since $\{x_i\}$ is a basis for X, T is one-to-one and onto. Because $\|Ty\| \leqslant \|y\|$, T is bounded. If $y_p = \{\alpha_{p_i}\}$ is a Cauchy sequence in Y, due to the completeness of X, Ty_p converges to an element x in X. From the definition of norm in Y it follows directly that $|\alpha_{p_i} - \alpha_{q_i}| \|x_i\| = \|(\alpha_{p_i} - \alpha_{q_i})x_i\| \leqslant \|y_p - y_q\|$. Hence there is a sequence $\{\alpha_i\} \subset \Phi$ such that $\lim_p \alpha_{p_i} = \alpha_i$ for all i and the rest of the proof follows exactly the lines of the rest of the proof for Theorem 1.3.

Lemma 3. *If $\{\alpha_i\} \in Y$, and $\{\beta_i\}$ is a sequence such that $|\beta_i| \leqslant |\alpha_i|$ for all i, then $\{\beta_i\}$ is in Y and $\|\{\beta_i\}\| \leqslant \|\{\alpha_i\}\|$.*

Proof. Since it may happen that $\alpha_i = 0$ for some i, we use the convention that $0/0 = 0$. Then $\|\{\beta_i\}\| = \sup\left\{\left\|\sum_{i \in \mu} \gamma_i \beta_i x_i\right\| \,\middle|\, \{\gamma_i\} \in S, \mu \in \Sigma\right\}$
$= \sup\left\{\left\|\sum_{i \in \mu} \gamma_i (\beta_i/\alpha_i)\alpha_i x_i\right\| \,\middle|\, \{\gamma_i\} \in S, \mu \in \Sigma\right\} \leqslant \sup\left\{\left\|\sum_{i \in \mu} \gamma_i \alpha_i x_i\right\| \,\middle|\, \{\gamma_i\} \in S, \mu \in \Sigma\right\}$
$= \|\{\alpha_i\}\|$. Finally, if $\lim_{m,n} \sum_{i=m}^{n} \alpha_i x_i = 0$, it follows from Lemma 2 and the

above estimate that $\lim_{m,n} \sum_{i=m}^{n} \beta_i x_i = 0$. By completeness of X, $\{\beta_i\}$ then is in Y and we have the lemma.

Based on these two lemmas we obtain

Theorem 4. *Let X^* be separable. Then each unconditional basis for X is shrinking.*

Proof. We suppose that an unconditional basis $\{x_i\}$ for X is not shrinking. Then there exists an $\varepsilon > 0$, an x^* in X^* of norm one, a sequence $\{y_i\}$ in X and a sequence $\{m_i\}$ of strictly increasing indices such that $\|y_i\| = 1$, $x^*(y_i) > \varepsilon$ and $y_i \in \mathrm{sp}\{x_{m_i+1}, \ldots, x_{m_{i+1}}\}$ (If the basis is not shrinking, then there is an $\varepsilon > 0$ and $0 \neq x^* \in X^*$ such that for every given index p there is an $x \notin \mathrm{sp}\{x_1, \ldots, x_p\}$ in X with $\|x\| = 1$ and $|x^*(x)| \geqslant 2\varepsilon$. Since $\{x_i\}$ is a basis for X, there is a $q > p$ with $\|x - U_q x\| < \varepsilon / \|x^*\|$, where U_q is an expansion operator of $\{x_i\}$. Hence with $y = [|x^*(U_q x)| / x^*(U_q x)] U_q x$ one obtains $x^*(y) = |x^*(U_q x)| \geqslant |x^*(x)| - \|x - U_q x\| > \varepsilon$, where now $y \in \mathrm{sp}\{x_{p+1}, \ldots, x_q\}$ and $\|y\| > 1 - \varepsilon$). Thus there is a sequence $\{\alpha_i\}$ in Φ such that $y_i = \sum_{j=m_i+1}^{m_{i+1}} \alpha_j x_j$. Now, let $\{\beta_i\}$ be any sequence in Φ. Using Lemmas 2 and 3 we obtain

$$\left\| \sum_{i \leqslant n} \beta_i y_i \right\| = \left\| \sum_{i \leqslant n} \beta_i \sum_{j=m_i+1}^{m_{i+1}} \alpha_j x_j \right\| \geqslant M \left\| \sum_{i \leqslant n} |\beta_i| \sum_{j=m_i+1}^{m_{i+1}} \alpha_j x_j \right\|$$

$$= M \left\| \sum_{i \leqslant n} |\beta_i| y_i \right\| \geqslant M \sum_{i \leqslant n} |\beta_i| x^*(y_i) > \varepsilon M \sum_{i \leqslant n} |\beta_i|,$$

where M is a constant, $0 < M \leqslant 1$. On the other hand, it follows immediately that $\left\| \sum_{i \leqslant n} \beta_i y_i \right\| \leqslant \sum_{i \leqslant n} |\beta_i| \|y_i\| = \sum_{i \leqslant n} |\beta_i|$ and the linear transformation T on l_1 (over the field Φ) to X, defined by $T\{\gamma_i\} = \lim_n \sum_{i \leqslant n} \gamma_i y_i$, $\{\gamma_i\} \in l_1$, is bounded. Since $\|T\{\gamma_i\}\| \geqslant \varepsilon M \|\{\gamma_i\}\|$, $\{\gamma_i\} \in l_1$, T^{-1} is also bounded so that T is a topological isomorphism of l_1 onto $T(l_1) \subset X$ (I.2.15). Hence l_1^* is topologically isomorphic with $T(l_1)^*$ (I.3.25), which is again topologically isomorphic with the factor space $X^*/T(l_1)^\perp$ (I.3.8). But this is impossible since X^* is separable and l_1^* is not (we observe that if $\{y_j^*\}$ is a countable dense set in X^*, so is $\{y_j^* + T(l_1)^\perp\}$ in $X^*/T(l_1)^\perp$). This contradiction leads to the conclusion that $\{x_i\}$ is shrinking and the theorem is proved.

Theorem 5. *If $\{x_i\}$ is an unconditional basis for X and X is weakly sequentially complete, then $\{x_i\}$ is boundedly complete.*

Proof. All spaces occurring in the proof are supposed to be over the same field. We use the facts that c_0^{**} is isometrically isomorphic with l_∞

(I.4.a and c), that c_0 is separable (I.4.a), l_∞ is not separable (I.4.c) and that $J(c_0)$ is weak* dense in c_0^{**} (I.3.22). Consequently, for every y^{**} in c_0^{**} there is a sequence $\{y_i\}$ in c_0 such that $y^{**}(x^*)=\lim_i x^*(y_i)$, $x^* \in c_0^*$. If c_0 would be weakly sequentially complete, then y^{**} would be in $J(c_0)$, i.e. c_0 would be isometrically isomorphic to l_∞, which is impossible. Thus c_0 is not weakly sequentially complete. Since by (I.3.15), every weakly converging sequence $\{y_i\}$ in the weakly sequentially complete space X converges weakly to an element in $\overline{\mathrm{sp}}\{y_i\}$, every closed linear subspace of X is itself weakly sequentially complete, hence can not be topologically isomorphic with c_0 (cf. proof of Corollary V.3.2).

Now, if $\{x_i\}$ is *not* boundedly complete, then, as we will demonstrate, there exists a subspace of X which is topologically isomorphic to c_0, and this contradiction shows that $\{x_i\}$ must be boundedly complete. In order to show this, we suppose the existence of a sequence $\{\alpha_i\}$ in Φ such that $\sup\left\{\left\|\sum_{i\in\mu}\alpha_i x_i\right\| \Big| \mu\in\Sigma\right\}\leqslant 1$ and such that $\sum_{i\leqslant n}\alpha_i x_i$ does not converge to any element of X. Thus, the series is not Cauchy in X and there exists an $\varepsilon>0$ and sequences of integers $\{n_i\}$ and $\{m_i\}$ such that $n_i\leqslant m_i<n_{i+1}$ for all i and $\left\|\sum_{j=n_i}^{m_i}\alpha_i x_i\right\|\geqslant\varepsilon$ for each i. Next, let Z_0 be the set of all sequences $\beta=\{\beta_i\}$ in c_0 which have only a finite number of non-vanishing elements and let $T:Z_0\to X$ be defined by $T\beta=\lim_n \sum_{i\leqslant n}\beta_i\sum_{j=n_i}^{m_i}\alpha_j x_j$, $\beta\in Z_0$. By Lemma 2 and 3 there is a constant $M>0$ such that

$$\|T\beta\|=\left\|\sum_i\beta_i\sum_{j=n_i}^{m_i}\alpha_j x_j\right\|\leqslant M\left\|\sum_i\|\beta\|\sum_{j=n_i}^{m_i}\alpha_j x_j\right\|$$

$$\leqslant M\|\beta\|\sup\left\{\left\|\sum_{j\in\mu}\alpha_j x_j\right\| \Big| \mu\in\Sigma\right\}\leqslant M\|\beta\|.$$

Therefore, $\|T\|\leqslant M$. By (I.2.12), since Z_0 is dense in c_0, T has a unique continuous linear extension, T', on the whole space c_0. From the same lemmas we obtain also

$$\|T'\beta\|=\left\|\lim_n\sum_{i\leqslant n}\beta_i\sum_{j=n_i}^{m_i}\alpha_j x_j\right\|\geqslant M^{-1}\sup\left\{\left\|\sum_{i=1}^{\infty}\delta_{ik}\beta_i\sum_{j=n_i}^{m_i}\alpha_j x_j\right\| \Big| k=1,2,\ldots\right\}$$

$$=M^{-1}\sup\left\{|\beta_k|\left\|\sum_{j=n_k}^{m_k}\alpha_j x_j\right\| \Big| k=1,2,\ldots\right\}\geqslant\varepsilon M^{-1}\|\beta\|,\qquad \beta\in c_0.$$

By (I.2.15) this estimate implies that T' has a bounded inverse. Thus c_0 is topologically isomorphic to $T'(c_0)\subset X$ and the proof of the theorem is complete.

Theorem 6. *If X^* is separable and $\{x_i, x_i^*\}$ is an unconditional basis for X, then $\{x_i^*, J x_i\}$ is an unconditional basis for X^*.*

Proof. By Theorem 4, $\{x_i, x_i^*\}$ is a shrinking basis for X. Theorem 3.9 and Corollary 3.5 imply that $\{x_i^*, J x_i\}$ is a boundedly complete basis for X^*. Let then $\{m_i\}$ be any indefinitely increasing sequence of integers. Since $\{x_i\}$ is unconditional the subseries $\sum_{i \leqslant n} x_{m_i}^*(x) x_{m_i}$ converges in X for

each x in X. Thus $\sup_n \left| \sum_{i \leqslant n} J x_{m_i}(x^*) x_{m_i}^*(x) \right| < \infty$, $x \in X$, $x^* \in X^*$. As a con-

sequence of Theorem I.3.14 it follows that $\sup_n \left\| \sum_{i \leqslant n} J x_{m_i}(x^*) x_{m_i}^* \right\| < \infty$,

$x^* \in X^*$. But since $\{x_i^*\}$ is a boundedly complete basis for X^* we infer that for every x^* in X^* the subseries $\sum_{i \leqslant n} J x_{m_i}(x^*) x_{m_i}^*$ converges in X^*.

Because $\{m_i\}$ was chosen arbitrarily, the expansion $\sum_{i \leqslant n} J x_i(x^*) x_i^*$ for x^*

in the basis $\{x_i^*\}$ converges unconditionally (II.1.3) and this finishes the proof.

Combining the results of Theorem 4.6, 3.9 and 4.4 we get immediately

Corollary 7. *If $\{x_i, x_i^*\}$ is an unconditional basis for X and if X^* is separable, then $\{x_i^*, J x_i\}$ is an unconditional boundedly complete basis for X^*.*

Corollary 7 applies in the special instance where X is reflexive, since then the separability of X implies the separability of X^* (I.3.11).

Theorem 8. *If $\{x_i, x_i^*\}$ is an unconditional basis for X and X^* is weakly sequentially complete, then $\{x_i^*\}$ is an unconditional basis for X^*.*

Proof. Let Σ be the set of all subsets of the set of all positive integers. As a consequence of Riemann's theorem (II.2.2), $\sum_{i=1}^{\infty} |x^*(x_i) x_i^*(x)|$ is finite

for each x in X and x^* in X^*. Therefore, $\sup \left\{ \left| \sum_{i \in \mu} x^*(x_i) x_i^*(x) \right| \middle| \mu \in \Sigma \right\} < \infty$.

(I.3.14) then implies that $\sup \left\{ \left\| \sum_{i \in \mu} x^*(x_i) x_i^* \right\| \middle| \mu \in \Sigma \right\} < \infty$, hence that

$\sup_\infty \left\{ \left| x^{**} \left(\sum_{i \in \mu} x^*(x_i) x_i^* \right) \right| \middle| \mu \in \Sigma \right\} < \infty$ for every x^{**} in X^{**}. This shows that

$\sum_{i=1}^{\infty} x^{**}[x^*(x_i) x_i^*]$ is absolutely convergent for all x^{**} in X^{**}. Now,

since X^* is weakly sequentially complete, the series $\sum_{i=1}^{\infty} x^*(x_i) x_i^*$ is sub-

series convergent in the weak topology of X^*. But due to the Orlicz-Pettis theorem, the series is strongly subseries convergent and hence unconditionally convergent in X^*. Moreover, the limit element must be x^*, since the series converges to x^* in the weak* topology of X^* (Theorem

2.7). Since $x^*(x_i) = J x_i(x^*)$, we have proved that $\{x_i^*, J x_i\}$ is an unconditional basis for X^*.

Corollary 9. (KARLIN) *There is no unconditional basis for the space* $C[0,1]$.

Proof. $C^*[0,1]$ is weakly sequentially complete (I.4.d). The corollary then follows from the preceding theorem since $C^*[0,1]$, again by (I.4.d) is non-separable.

5. Absolutely Convergent Bases and Uniform Bases

Definition 1. *A basis* $\{x_i, x_i^*\}$ *for* X *is called absolutely convergent if* $\sum\limits_{i=1}^{\infty} x_i^*(x) x_i$ *converges absolutely for every* x *in* X.

It is clear that every absolutely convergent basis is automatically an unconditional basis.

Theorem 2. *If* X *possesses an absolutely convergent basis, then* X *is topologically isomorphic to* l_1.

Proof. Let $\{x_i, x_i^*\}$ be an absolutely convergent basis for X. We define the mapping $T: X \to l_1$ by $T x = \{x_i^*(x) \|x_i\|\}$, $x \in X$. Clearly, T is linear. Let $\{y_j\}$ be a sequence in X such that $\lim\limits_{j} y_j = y \in X$ and such that $T y_j$ converges to some element $\{\alpha_i\}$ in l_1. Then $\lim\limits_{j} \sum\limits_{i=1}^{\infty} |x_i^*(y_j)| \|x_i\| - \alpha_i| = 0$. But this implies that $T y = \{\alpha_i\}$, since $x_i^*(y) \|x_i\| = x_i^* \left(\lim\limits_{j} y_j \right) \|x_i\| = \lim\limits_{j} x_i^*(y_j) \|x_i\|$ $= \alpha_i$ for all i. Thus T is closed and by (I.2.10) it follows that T is bounded. Furthermore, T is onto, because for every $\{\alpha_i\}$ in l_1, $x = \sum\limits_{i=1}^{\infty} (\alpha_i/\|x_i\|) x_i$ is in X and $T x = \{\alpha_i\}$. On the other hand we obtain that $\|x\|$ $= \left\| \lim\limits_{n} \sum\limits_{i \leq n} x_i^*(x) x_i \right\| \leq \sum\limits_{i=1}^{\infty} |x_i^*(x)| \|x_i\| = \|T x\|$ for each x in X. As a result of this (I.2.15), X is topologically isomorphic with l_1.

Definition 3. *A basis* $\{x_i, x_i^*\}$ *for* X *is uniform if* $\sum\limits_{i=1}^{\infty} x_i^*(x) x_i$ *converges uniformly for all* x *in the unit ball of* X.

Lemma 4. *If* X *is* n-dimensional, then there exists a basis $\{x_1, \dots, x_n\}$ for X with associated biorthogonal sequence of coefficient functionals $\{x_i^*, \dots, x_n^*\}$ in X^*.

Proof. X has a Hamel basis, since each linearly independent set of n elements $\{x_1,\ldots,x_n\}$ in X forms such a basis. The existence of a biorthogonal sequence of coefficient functionals $\{x_1^*,\ldots,x_n^*\}$ follows from Theorem III.2.4 or directly from the following consideration: Due to (I.1.17) there is a topological isomorphism T of X onto E^n. Let $\{y_i\}$ be an orthonormal basis for E^n. Then $x = T^{-1}\left[\sum_{i=1}^{n}(Tx,y_i)y_i\right] = \sum_{i=1}^{n}(Tx,y_i)T^{-1}y_i$. Since the set $\{T^{-1}y_i\}$ is linearly independent in X, the set $\{(Tx,y_i)\}$ is unique for each x in X. Moreover $x_i^*:X\to\Phi$, defined by $x_i^*(x)=(Tx,y_i)$ is linear and, by $|x_i^*(x)|\leqslant\|T\|\,\|y_i\|\,\|x\|$, bounded. This implies that $\{T^{-1}y_i,x_i^*\}$ is a basis for X and we are done.

Theorem 5. *There exists a uniform basis for X if and only if X is finite dimensional.*

Proof. Let X be finite dimensional, say of dimension n. Then by the lemma there exists a basis $\{x_1,\ldots,x_n,x_1^*,\ldots,x_n^*\}$ for X. Since $x = \sum_{i=1}^{n}x_i^*(x)x_i$ for all x in X, the basis is obviously uniform. Conversely, the assumption of a uniform basis $\{x_i,x_i^*\}$ for X with expansion operators U_n, $n=1,2,\ldots$, implies that $\lim_n\|I-U_n\| = \lim_n\sup\left\{\left\|x-\sum_{i\leqslant n}x_i^*(x)x_i\right\|\,\big|\,\|x\|\leqslant 1\right\} = 0$. Since each U_n is bounded and with finite dimensional range, by (I.1.12), each U_n is compact. It thus follows (I.2.17) that I is also compact, a fact which is possible only in finite dimensional spaces (I.1.12). This completes the proof of the theorem.

6. T-Bases

We shall now treat summability methods and their relevance to Banach spaces. Use is made of infinite matrices with special properties which are known as consistent or Toeplitz matrices. We begin with.

Definition 1. *Let $T=\{t_{ij}\}$ be an infinite matrix with entries in Φ. Then a sequence $\{x_i\}$ is T-limitable to x in X if $y_i = \lim_n\sum_{j\leqslant n}t_{ij}x_j$ exists for all i and y_i converges to x in X.*

Definition 2. *An infinite matrix T is consistent if and only if, whenever a sequence in X converges to x in X, the sequence is also T-limitable to x in X.*

Theorem 3. *An infinite matrix $T=\{t_{ij}\}$ is consistent if and only if the following conditions are satisfied:*

(i) $$\sum_{j=1}^{\infty} |t_{ij}| \leqslant M < \infty, \qquad i \geqslant 1,$$

(ii) $$\lim_i t_{ij} = 0, \qquad j \geqslant 1 \quad and$$

(iii) $$\lim \sum_{j=1}^{\infty} t_{ij} = 1.$$

Proof. Necessity. Let $\{\alpha_j\}$ be a convergent sequence in Φ with limit α_0 and let $x \in X$ and $x^* \in X^*$ be such that $x^*(x) = 1$. Then, since T is consistent and since $\alpha_i x$ converges to $\alpha_0 x$, $y_i = \sum_{j=1}^{\infty} t_{ij} \alpha_j x$ also converges to $\alpha_0 x$. Hence $\sum_{j=1}^{\infty} t_{ij} \alpha_j = x^* \left(\sum_{j=1}^{\infty} t_{ij} \alpha_j x \right) = x^*(y_i)$ converges to α_0. This shows that T is consistent in the numerical case.

Therefore we consider the linear functionals τ_i on the space c_0, given by $\tau_i(\alpha) = \sum_{j=1}^{\infty} t_{ij} \alpha_j, \alpha = \{\alpha_i\} \in c_0$. We first fix i and show that each τ_i is bounded on c_0. Assuming the contrary would imply that τ_i is not in c_0^*, hence that $\sum_{j=1}^{\infty} |t_{ij}| = \infty$ (I.4.a). In other words the sequence $\left\{ \sum_{j \leqslant n} |t_{ij}| \right\}$ would be not Cauchy in $\mathbb{R}$. Hence there would be an $\varepsilon > 0$ and an increasing sequence $\{m_n\}$ such that $\sum_{j=m_n+1}^{m_{n+1}} |t_{ij}| > \varepsilon$. We then can take $\alpha_i = \bar{t}_{ij}/(n|t_{ij}|)$ if $m_n < i \leqslant m_{n+1}$ and $t_{ij} \neq 0$, else $\alpha_i = 0$. Clearly $\{\alpha_i\}$, so chosen, is in c_0. But $\sum_{j=1}^{\infty} t_{ij} \alpha_j \geqslant \sum_{n=1}^{\infty} (1/n) \sum_{j=m_n+1}^{m_{n+1}} |t_{ij}| \geqslant \varepsilon \sum_{n=1}^{\infty} (1/n) = \infty$ which is a contradiction. Next, since $\lim_i \tau_i(\alpha)$ exists for every α, $\sup_i |\tau_i(\alpha)| < \infty, \alpha \in c_0$, and the uniform boundedness principle (I.3.14) implies $\sup_i \|\tau_i\| \leqslant M < \infty$. Thus $\left| \sum_{j=1}^{\infty} t_{ij} \alpha_j \right| \leqslant M \sup_j |\alpha_j|, i = 1, 2, \ldots$ and choosing $\alpha_j = \bar{t}_{ij}/|t_{ij}|$ for $j \leqslant n$ if $t_{ij} \neq 0$ and $\alpha_j = 0$ else, we obtain $\sum_{j=1}^{n} |t_{ij}| \leqslant M, n = 1, 2, \ldots$, from which (i) follows immediately. Finally, let $\alpha^{(n)} = \{\delta_{nj}\}$ and $\alpha^{(0)} = \{1, 1, \ldots\}$. Then $0 = \lim_i \alpha_i^{(n)} = \lim_i \sum_{j=1}^{\infty} t_{ij} \alpha_j^{(n)} = \lim_i t_{in}, n = 1, 2, \ldots$, which is (ii) and $1 = \lim_i \alpha_i^{(0)} = \lim_i \sum_{j=1}^{\infty} t_{ij} \alpha_j^{(0)} = \lim_i \sum_{j=1}^{\infty} t_{ij}$, which is (iii). Sufficiency. It is convenient to choose a sequence $\{x_i\}$ in X, converging to x in X, which is such that $\sup_i \|x_i\| \leqslant 1$.

By condition (ii) and (iii) and the convergence of $\{x_i\}$ there exists for every $\varepsilon > 0$ an n and an i_0 depending on n, such that for $i \geqslant i_0$,

$$\sum_{j<n} |t_{ij}| < \varepsilon/6, \left|1 - \sum_{j=1}^{\infty} t_{ij}\right| < \varepsilon/3 \text{ and } \|x - x_j\| < \varepsilon/(3M), j \geqslant n. \text{ Hence, due to}$$

(i), $\left\|x - \sum_{j=1}^{\infty} t_{ij}x_j\right\| \leqslant \left\|x\left(1 - \sum_{j=1}^{\infty} t_{ij}\right)\right\| + \left\|\sum_{j<n} t_{ij}(x - x_j)\right\| + \left\|\sum_{j=n}^{\infty} t_{ij}(x - x_j)\right\|$

$< \varepsilon/3 + 2\varepsilon/6 + M\varepsilon/(3M) = \varepsilon, i \geqslant i_0$. This shows that the sequence $\{x_i\}$ is T-limitable to x and the proof of the theorem is finished.

Definition 4. *Let T be an infinite matrix with entries in Φ. Then a sequence $\{x_i\}$ in X is called a T-basis for X if for every x in X there is a unique coefficient sequence $\{\alpha_i\}$ in Φ such that the sequence $\left\{\sum\limits_{i \leqslant n} \alpha_i x_i\right\}$ is T-limitable to x.*

We observe that the uniqueness implies that each x_i is non-zero and that each basis for X is a T-basis for X with $T = \{\delta_{ij}\}$.

Theorem 5. *Let $T = \{t_{mn}\}$ be a consistent invertible lower triangular matrix and let $\{x_i\}$ be a T-basis for X. Then there is a unique sequence $\{x_i^*\}$ in X^* such that $x_i^*(x) = \alpha_i, x \in X$ and that $x_i^*(x_j) = \delta_{ij}$.*

Proof. Let Y be the linear space of those sequences $y = \{\alpha_i\}$ in Φ for which the partial sums $\sum\limits_{i \leqslant n} \alpha_i x_i$ are T-limitable to an element x of X and let $z_p = \sum\limits_{n=1}^{\infty} t_{pn} \sum\limits_{i \leqslant n} \alpha_i x_i = \sum\limits_{n \leqslant p} t_{pn} \sum\limits_{i \leqslant n} \alpha_i x_i$. We define the norm in Y by $\|y\| = \sup\limits_{p} \|z_p\|, y \in Y$ and we will show that the normed linear space Y is complete.

If $T^{-1} = \{u_{mn}\}$, which is again a lower triangular matrix, we note that $\sum\limits_{i \leqslant n} \alpha_i x_i = \sum\limits_{p=1}^{\infty} u_{np} z_p = \sum\limits_{p \leqslant n} u_{np} z_p$. Now, let $\{y_n\}$ be a Cauchy sequence in Y, where $y_n = \{\alpha_{ni}\}$. It is then clear that $\lim\limits_{n} \alpha_{ni} = \alpha_i$ exists and we will prove that $y = \{\alpha_i\} \in Y$ and that $y = \lim\limits_{n} y_n$ in Y. To this end let $z_{pn} = \sum\limits_{j \leqslant p} t_{pj} \sum\limits_{i \leqslant j} \alpha_{ni} x_i$. Obviously, $\|z_p - z_q\| \leqslant \|z_p - z_{pn}\| + \|z_{pn} + z_{qn}\|$ $+ \|z_{qn} + z_q\|$ and $\|z_{pn} - z_{pm}\| \leqslant \|y_n - y_m\|$. Taking the limit on m we get for every $\varepsilon > 0$ an n such that $\|z_{pn} - z_p\| \leqslant \varepsilon/3$, independently of p. Since $y_n \in Y$ we can choose an r, depending on n, such that for $p, q \geqslant r$, $\|z_{pn} - z_{qn}\| < \varepsilon/3$. This implies $\|z_p - z_q\| < \varepsilon, p, q \geqslant r$ so that $y \in Y$. That Y is complete finally follows from $\|y - y_n\| = \sup\limits_{p} \|z_p - z_{pn}\|$ which, by the above, can be made arbitrarily small.

Next, let $S: Y \rightarrow X$ be the one-to-one linear transformation defined by $Sy = x$. Evidently, $\|S\| \leqslant 1$ so that, since S is onto, S^{-1} is also bounded

(I.2.6). Defining linear functionals x_i^* on X by $x_i^*(x)=\alpha_i$, we obtain
$|x_1^*(x)|=|\alpha_1|=\|\alpha_1 x_1\|/\|x_1\| \leqslant \|x_1\|^{-1}\|u_{11} z_1\| \leqslant \|x_1\|^{-1}|u_{11}|\|y\|$
$\leqslant \|x_1\|^{-1}|u_{11}|\|S^{-1}\|\|x\|$. In a similar way one has for $i>1$, $|x_i^*(x)|$
$= \left\|\sum_{j\leqslant i}\alpha_j x_j - \sum_{j<i}\alpha_j x_j\right\|/\|x_i\| \leqslant 2\|x_i\|^{-1}\sup\left\{\left\|\sum_{p\leqslant k}u_{kp}z_p\right\| \,\Big|\, i-1\leqslant k\leqslant i\right\}$
$\leqslant 2\|x_i\|^{-1}K_i\|S^{-1}\|\|x\|$, where $K_i=\sup\left\{\sum_{p\leqslant k}|u_{kp}| \,\Big|\, i-1\leqslant k\leqslant i\right\}<\infty$.

Thus each x_i^* is an element of X^*. Finally, the consistency of T
implies that $\lim\limits_i \sum\limits_{j=n}^{\infty} t_{ij}=1$ for all n, hence that $\sum\limits_{i\leqslant m}\delta_{ni}x_i$ is T-limitable
to x_n which demonstrates that $x_i^*(x_n)=\delta_{in}$ and the theorem is proved.

Theorem 6. *Let $\{x_i\}$ be a total sequence in X with biorthogonal set $\{x_i^*\}$
in X^* such that for some consistent matrix $T=\{t_{ij}\}$, $\lim\limits_p \sum\limits_{i\leqslant p} t_{ni} \sum\limits_{j\leqslant i} x_j^*(x)x_j$
exists for $n=1,2,\ldots$ and $\sup\limits_n\left\|\sum\limits_{i=1}^{\infty} t_{ni} \sum\limits_{j\leqslant i} x_j^*(x)x_j\right\|<\infty$ for each x in X.
Then $\{x_i\}$ is a T-basis for X and $\{x_i^*(x)\}$ is the coefficient sequence corre-
sponding to X.*

Proof. Let $T_n: X\to X$, $n=1,2,\ldots$ be defined by $T_n x = \sum\limits_{i=1}^{\infty} t_{ni} \sum\limits_{j\leqslant i} x_j^*(x)x_j$.
The hypothesis and the principle of uniform boundedness (I.3.14) then
imply that $\sup\limits_n\|T_n\| \leqslant M<\infty$. Since $\overline{\mathrm{sp}}\{x_i\}=X$ we have for each x in X
and every $\varepsilon>0$ an m and an elementy y_m in $\mathrm{sp}\{x_i|i\leqslant m\}$ such that $\|x-y_m\|$
$<\varepsilon$. On the other hand, since T is consistent and since $\sum\limits_{j\leqslant i} x_i^*(y_m)x_i=y_m$ for
all $i\geqslant m$, there is an n_ε such that $\|y_m - T_n y_m\| = \left\|y_m - \sum\limits_{i=1}^{\infty} t_{ni} \sum\limits_{j\leqslant i} x_j(y_m)x_j\right\|$
$<\varepsilon$, $n\geqslant n_\varepsilon$. But then, $\|x-T_n x\| \leqslant \|x-y_m\| + \|y_m - T_n y_m\| + \|T_n(y_m-x)\|$
$\leqslant(2+M)\varepsilon$ if $n\geqslant n_\varepsilon$ so that $\sum\limits_{j\leqslant i} x_i^*(x)x_j$ is T-limitable to x. It now remains
to show that the coefficient sequence $\{x_i^*(x)\}$ for x is unique. This
follows from the assumption $\lim\limits_n \sum\limits_{i=1}^{\infty} t_{ni} \sum\limits_{j\leqslant i} \alpha_j x_j=0$ for some sequence
$\{\alpha_i\}$ in Φ, and by multiplication of this equation with x_m^*. Then
$0=x_m^*\left(\lim\limits_n \sum\limits_{i=1}^{\infty} t_{ni} \sum\limits_{j\leqslant i} \alpha_j x_j\right)=\alpha_m\lim\limits_n \sum\limits_{i=m}^{\infty} t_{ni}=\alpha_m\left(\lim\limits_n \sum\limits_{i=1}^{\infty} t_{ni}-\lim\limits_n \sum\limits_{i<m} t_{ni}\right)=\alpha_m$.
Hence $\{\alpha_i\}$ consists of zeros and the sequence $\{x_i^*(x)\}$ is unique.

Example. It is easy to see that the matrix $T=\{t_{ij}\}$, given by

$$t_{ij}=\begin{cases}1/i, & j\leqslant i,\\ 0, & j>i,\end{cases}$$

is consistent. The corresponding T-basis for X is called a *Cesaro basis*
for X.

7. Bases for Special Spaces

Though it is still an open question wether there exists a basis for *every* separable Banach space, bases are known for almost all of the usually encountered examples of infinitely dimensional separable Banach spaces. We begin with the spaces which are the easiest to treat, the spaces of sequences, c_0 and l_p, $1 \leqslant p < \infty$. We use the symbol δ_i for the sequence $\{\delta_{ij}\}$ in Φ, where δ_{ij} is, as usual, Kronecker's delta.

Theorem 1. $\{\delta_i\}$ *is a monotone unconditional and non-boundedly complete basis for c_0 with biorthogonal sequence $\{\delta_i\}$ in $c_0^* = l_1$.*

Proof. If $\alpha = \{\alpha_i\}$ belongs to c_0, then it is clear that $\alpha = \lim\limits_n \sum\limits_{i \leqslant n} \alpha_i \delta_i$. As an immediate consequence of the obvious equation $\alpha_i = \delta_i^*(\alpha)$, where $\delta_i^* \in c_0^* = l_1$ is given by $\delta_i^* = \delta_i$, and Theorem 2.4, $\{\delta_i\}$ is a basis for c_0. The basis is unconditional since the definition $\|\alpha\| = \sup\limits_i |\alpha_i|$ of the norm in c_0 implies that the expansion for α is subseries, hence unconditionally convergent (II.1.3). That the basis is monotone is immediately clear from the definition of norm in c_0. The following counter-example shows that the basis is not boundedly complete: The sequence $\alpha = \{1,1,\ldots\}$ in Φ is such that $\sup\limits_n \left\| \sum\limits_{i \leqslant n} \alpha_i \delta_i \right\| = 1$. But $\sum\limits_{i \leqslant n} \alpha_i \delta_i$ is not convergent in c_0 and this contradiction concludes the proof of the theorem.

Theorem 2. *Let $x_{ij} = 1$ for $j \leqslant i$ and $x_{ij} = 0$ for $j > i$. Then with $x_i = \{x_{ij}\}$, $\{x_i\}$ is a basis for c_0 which is not unconditional.*

Proof. It is easy to verify that the sequence $\{x_i^*\}$ in l_1, given by $x_i^* = \delta_i - \delta_{i+1}$, is biorthogonal to $\{x_i\}$. Since $\overline{\mathrm{sp}}\{x_i\} = c_0$ and since
$$\sup\limits_n \|U_n \alpha\| = \sup\limits_n \left\| \sum\limits_{i \leqslant n} x_i^*(\alpha) x_i \right\| = \sup\limits_n \left\| \sum\limits_{i \leqslant n} (\alpha_i - \alpha_{i+1}) x_i \right\| = \sup\limits_n \sup\{ |\alpha_j - \alpha_n| \, | \, j < n \}$$
$\leqslant 2 \sup\limits_n |\alpha_n| = 2\|\alpha\|$, Corollary 2.3 implies that $\{x_i\}$ is a basis for c_0. We use an example to show that $\{x_i\}$ is not unconditional. Let α be an element of c_0 given by $\alpha = \{(-1)^{i+1}/i\}$. We then take the subseries of all odd terms in the expansion of α:

$$\sum\limits_{i=1, i \text{ odd}}^{\infty} x_i^*(\alpha) x_i = \sum\limits_{i=1, i \text{ odd}}^{\infty} (\alpha_i - \alpha_{i+1}) x_i$$

$$= \left\{ \sum\limits_{i=2j-1}^{\infty} (-1)^{i+1} \alpha_i \right\}$$

$$= \left\{ \sum\limits_{i=2j-1}^{\infty} i^{-1} \right\}.$$

Finally, from the divergence of the series $1+\frac{1}{2}+\frac{1}{3}+\ldots$ we conclude that the series expansion for α is not subseries, hence not unconditional convergent (II.1.3) and the proof of the theorem is complete.

Theorem 3. $\{\delta_i\}$ *is a monotone unconditional retro-basis for* l_p, $1\leqslant p<\infty$. *The basis is absolutely convergent for* $p=1$ *but not for* $p>1$.

Proof. The biorthogonal sequence to $\{\delta_i\}$ is $\{\delta_i^*\}$, $\delta_i^* \in l_p^* = l_q$ ($1/p+1/q = 1$) given by $\delta_i^* = \delta_i$. Since for $\alpha \in l_p$, $\delta_i^*(\alpha) = \alpha_i$, $\sum_{i \leqslant n} \delta_i^*(\alpha)\delta_i$ converges to α in the topology of each of the spaces l_p, $1\leqslant p<\infty$. Therefore, the theorem is a consequence of Theorem 2.4 and $\{\delta_i\}$ is a retro-basis since for all i, δ_i is an element of c_0 as well as of l_q, $1<q<\infty$. To show that $\{\delta_i\}$ is unconditional, let $\{n_i\}$ be any increasing sequence of integers. Since l_p is complete and because $\left\| \sum_{n_i \in \mu} \delta_{n_i}^*(\alpha)\delta_{n_i} \right\| = \left(\sum_{n_i \in \mu} |\alpha_{n_i}|^p \right)^{1/p}$
$\leqslant \left(\sum_{i \in \mu} |\alpha|^p \right)^{1/p}$ for each α in l_p and each finite set of integers μ, the expansion for α is subseries convergent. Thus $\{\delta_i\}$ is unconditional for $1\leqslant p<\infty$ (II.1.3). Next, let $p>1$ and let p' be such that $1<p'<p$. Then $\alpha = \{i^{-p'/p}\}$ is an element of l_p because $\|\alpha\|^p = \sum_{i=1}^{\infty} (i^{-p'/p})^p = \sum_{i=1}^{\infty} i^{-p'} \leqslant 1 + \int_1^{\infty} t^{-p'} dt$
$= 1-(1-p')^{-1} < \infty$. But $\sum_{i \leqslant n} \|\delta_i^*(\alpha)\delta_i\| = \sum_{i \leqslant n} |\alpha_i| = \sum_{i \leqslant n} i^{-p'/p} \geqslant \int_1^n t^{-p'/p} dt$
$= (1-p'/p)^{-1}(n^{1-p'/p}-1)$ which diverges as $n \to \infty$, showing that in this case the convergence of the series expansion for α is not absolutely convergent. However, the basis is absolutely convergent for $p=1$ since $\sum_{i=1}^{\infty} \|\delta_i^*(\alpha)\delta_i\| = \sum_{i=1}^{\infty} |\alpha_i| = \|\alpha\|$. The monotony finally follows directly from the definition of the norm in l_p.

Theorem 4. *Let* $x_1 = \{1,0,0,\ldots\}$, $x_2 = \{-1,1,0,0,\ldots\}$, $x_3 = \{1,0,1,0,0,\ldots\}$, ... *Then* $\{x_i\}$ *is a weak* as well as an absolutely convergent basis for* l_1, *but* $\{x_i\}$ *is neither a retro-basis nor a weak* Schauder basis for* l_1.

Proof. It can be verified directly that $x_1^* = \{1,1,-1,1,-1,\ldots\}$ and $x_i^* = \delta_i$, $i>1$, are biorthogonal functionals in $l_1^* = l_\infty$. Let α be in l_1. Then

$$\left\| \alpha - \sum_{i \leqslant n} x_i^*(\alpha) x_i \right\| = \left\| \left\{ -\sum_{j=n+1}^{\infty} (\alpha_j)^j, 0,\ldots,0, \alpha_{n+1}, \alpha_{n+2}, \ldots \right\} \right\| \leqslant 2 \sum_{j=n+1}^{\infty} |\alpha_j|.$$

Since the last term converges to zero with n, Theorem 2.4 implies that $\{x_i\}$ is a basis for l_1. The basis is absolutely convergent since

$$\sum_{i=1}^{\infty} \|x_i^*(\alpha) x_i\| = \left| \alpha_1 + \sum_{j=2}^{\infty} (\alpha_j)^j \right| + 2 \sum_{j=2}^{\infty} |\alpha_j| \leqslant 3\|\alpha\|.$$

Finally, $\{x_i\}$ cannot be a retro-basis because x_1^* ist not in c_0. $\{x_i\}$ can not be a weak* Schauder basis for l_1 since then Theorem 2.7 would imply that $x_1^* \in c_0$ which is a contradiction. However, $\{x_i\}$ is a weak* basis for l_1 since it is a basis for l_1 and since $\lim_n \sum_{i \leqslant n} \alpha_i (x_i)_j = 0, j \geqslant 1$, implies $0 = \alpha_2 = \alpha_3 = \ldots$ and thus $\alpha_1 = 0$.

Recalling that χ_S denotes the characteristic function of a set S we have

Theorem 5. *The following sequence in $C[0,1]$ (Schauder's system) is a monotone basis for $C[0,1]$:*

$$x_0(t) = \chi_{[0,1]}(t),$$

$$x_1(t) = t \chi_{[0,1]}(t),$$

$$x_2(t) = x_1(2t) + \chi_{(0,1]}(2t-1) - x_1(2t-1),$$

$$x_{2^n+i}(t) = x_2(2^n t - i + 1), \quad i = 1,\ldots, \quad n = 1,2,3,\ldots .$$

Proof. Let $x \in C[0,1]$ be arbitrary and let $y_n \in C[0,1]$ for $n \geqslant 1$ be defined by $y_n(t) = x(t)$, $t = 0, 1/2^n, 2/2^n, 3/2^n, \ldots, 1$, variing linearly between these points. Since $x \in C[0,1]$ it is uniformly continuous (I.4.d) on $[0,1]$ and the polygonal functions y_n converge uniformly to x. Clearly, $y_n \in \overline{\mathrm{sp}}\{x_0, \ldots, x_{2^n}\}$. Hence $\{x_i\}$ is total in $C[0,1]$. Furthermore it is evident that $\left\| \sum_{i \leqslant n} \alpha_i x_i \right\| \leqslant \left\| \sum_{i \leqslant n+1} \alpha_i x_i \right\|$ for all $n \geqslant 1$ and arbitrary numbers $\alpha_1, \alpha_2, \ldots$ in Φ. Thus the theorem is a consequence of Theorem IV.1.5. The inequality also implies the monotony of the basis $\{x_i\}$.

Unfortunately, here (and in the following proof) we make use of a theorem which is proved later in chapter IV. But we feel that this is justified, since we like to have some applications in classical Banach spaces already now and it is desirable to have them all together.

Theorem 6. *Haar's system, which is given by*

$$x_1(t) = \chi_{[0,1]}(t),$$

$$x_{2^n+j}(t) = 2^{n/2} [\chi_{[0,1]}(2^{n+1} t - 2j + 2) - \chi_{(0,1]}(2^{n+1} t - 2j + 1)],$$

$$j = 1, \ldots, 2^n, \quad n = 0, 1, 2, \ldots,$$

is a monotone basis for $L_p[0,1]$, $1 \leqslant p < \infty$.

Proof. Let $x \in C[0,1]$ be arbitrary and let the function $y_i \in \mathrm{sp}\{x_j | j \leqslant 2^i\}$, $i \geqslant 0$ be defined by $y_i(t) = x(t)$, $t = j/2^i, j = 1, 2, \ldots, 2^i$. Since $x(t)$ is continuous, it is uniformly continuous (I.4.d). Hence the step functions y_i converge with i uniformly to x on $[0,1]$. Therefore, $\lim_i\|x-y_i\|=0$ and, since the subspace $C[0,1]$ of $L_p[0,1]$ is dense in $L_p[0,1]$, $1\leqslant p<\infty$ (I.4.8), Haar's system $\{x_i\}$ is total in $L_p[0,1]$. For $y\in\mathrm{sp}\{x_j|j<i\}$, $i=2,3,4,\ldots$ we have for all $\alpha\in\Phi$,

$$\int_{t_i^-}^{t_i^+}|y(t)+\alpha x_i(t)|^p dt = \frac{t_i^+-t_i^-}{2}\left[|y(t_i^+)+\alpha 2^{n/2}|^p+|y(t_i^+)-\alpha 2^{n/2}|^p\right]$$

$$\geqslant (t_i^+-t_i^-)|y(t_i^+)|^p = \int_{t_i^-}^{t_i^+}|y(t)|^p dt,$$

where $t_i^+=\sup\{t|t\in\mathrm{supp}(x_i)\}$ and $t_i^-=\inf\{t|t\in\mathrm{supp}(x_i)\}$. Thus, $\|y\|\leqslant\|y+\alpha x_i\|$ and according to Theorem IV.1.5, $\{x_i\}$ is a basis for $L_p[0,1]$ and this basis is obviously monotone. Since the functionals on $L_p[0,1]$, defined by $x_i^*(x)=\int_0^1 x(t)x_i(t)dt$, $i=1,2,\ldots$ are bounded and satisfy $x_i^*(x_j)=\delta_{ij}$ and since the associated biorthogonal sequence of a basis is unique, we notice that $\{x_i^*\}$ is just the biorthogonal sequence of $\{x_i\}$. This completes the proof of the theorem.

Theorem 7. *Let $\{x_i\}$ be the sequence of functions given by $x_0(t)=(2\pi)^{-\frac{1}{2}}$, $x_{2i-1}(t)=\pi^{-\frac{1}{2}}\sin(it)$, $x_{2i}(t)=\pi^{-\frac{1}{2}}\cos(it)$, $0\leqslant t\leqslant 2\pi$, $i=1,2,\ldots$. Then $\{x_i\}$, the trigonometrical system, is an unconditional basis for the Hilbert space $L_2[0,2\pi]$.*

Proof. It is easy to see that $\{x_i\}$ is an orthonormal system for $L_2[0,2\pi]$ (i.e. $(x_i,x_j)=\delta_{ij}$). Let the sequence $\{x_i^*\}$ in $L_2^*[0,2\pi]$ be defined by $x_i^*(x)=(x,x_i)$, $x\in L_2[0,2\pi]$. Since $\{x_i,x_i^*\}$ is evidently a biorthogonal system for $L_2[0,2\pi]$ and since, according to Theorem I.4.11, $\lim_n\sum_{i\leqslant n}x_i^*(x)x_i=x$ for each x in $L_2[0,2\pi]$, Theorem 2.4 implies that $\{x_i\}$ is a basis for $L_2[0,2\pi]$. Furthermore, it is possible to show that $\{x_i\}$ is unconditional. Let $\{n_i\}$ be any increasing sequence of integers. Then, since

$$\left\|\sum_{i=p}^q x_{n_i}^*(x)x_{n_i}\right\| = \left\|\sum_{i=p}^q (x,x_{n_i})x_{n_i}\right\| = \sum_{i=p}^q |(x,x_{n_i})|^2$$

$$\leqslant \sum_{i=n_p}^\infty |(x,x_i)|^2,$$ and since by Bessel's inequality, $\sum_{i=0}^\infty|(x,x_i)|\leqslant\|x\|^2$, the completeness of $L_2[0,2\pi]$ implies that the subseries $\sum_{i=1}^\infty (x,x_{n_i})x_{n_i}$ is convergent in $L[0,2\pi]$ for every x in $L_2[0,2\pi]$ and from (II.1.3) the assertion follows.

To proof the next theorem we make use of the *Rademacher functions* ψ_n, defined for $n=1,2,\ldots$ on the interval $[0,1]$ of $\mathbb{R}$ by $\psi_n(s)=\mathrm{sign}\sin(2^n\pi s)$, where $\mathrm{sign}\, s=-1,=0$ or $=1$, according to $s<0,=0$ or >0, respectively. Let $E_n^{\pm} = \{s | \psi_n(s) \gtrless 0\}$ and denote by $|E|$ the Lebesgue-measure of a measurable subset of $[0,1]$. It is then clear that

$$\int_0^1 \psi_n(s) \psi_m(s) ds = |E_n^+ \cap E_m^+| + |E_n^- \cap E_m^-| - |E_n^- \cap E_m^+| - |E_n^+ \cap E_m^-|$$

$$= (1 + \delta_{nm})(\tfrac{1}{4} + \tfrac{1}{4}) - (1 - \delta_{nm})(\tfrac{1}{4} + \tfrac{1}{4}) = \delta_{nm},$$

i.e. $\{\psi_n\}$ is an orthonormal system in the Hilbert space $L_2[0,1]$.

Theorem 8. *The sequence $\{x_i\}$ of Theorem 7 forms a basis for $L_p[0,2\pi]$, $1 < p < \infty$, which is not unconditional for $p \neq 2$.*

Proof. Let $1 < p < \infty$ and let $q = p/(p-1)$. First we assume $L_p[0,2\pi]$ to be real. We take in $L_p^*[0,2\pi]$ (which is by (I.4.6) isometrically isomorphic to $L_q[0,2\pi]$) a sequence $\{x_i^*\}$ defined by $x_i^*(x) = \int_0^{2\pi} x_i(t)x(t)dt$, $x \in L_p[0,2\pi]$, $i = 0,1,\dots$. Clearly, the system $\{x_i, x_i^*\}$ is biorthogonal and since by Theorem I.4.11, $\lim_n \left\| x - \sum_{i \leq n} x_i^*(x)x_i \right\| = 0$ for every x in $L_p[0,2\pi]$, as a consequence of Theorem 2.4, $\{x_i\}$ is a basis for $L_p[0,2\pi]$. First we restrict ourself to a fixed p with $1 < p < 2$. Now, if for every x in $L_p[0,2\pi]$ we would have $\sum_{i=0}^{\infty} |x_i^*(x)|^2 < \infty$, then since $\left\| \sum_{i=m}^n x_i^*(x)x_i \right\|_2^2$

$= \sum_{i=n}^m |x_i^*(x)|^2$ and since $L_2[0,2\pi]$ is complete, $\sum_{i \leq n} x_i^*(x)x_i$ converges to an element y in $L_2[0,2\pi]$ in the topology of $L_2[0,2\pi]$, where $\|f\|_2$ denotes the norm of f as an element of $L_2[0,2\pi]$. Due to Hölder's inequality, $\|f\|^p = \int_0^{2\pi} |f(t)|^p dt \leq \left[\int_0^{2\pi} |f(t)|^{p(2/p)} dt \right]^{p/2} \cdot (2\pi)^{1 - p/2} = (2\pi)^{1-p/2} \|f\|_2^p$

for all f in $L_p[0,2\pi]$. Hence y is in $L_p[0,2\pi]$ and $\sum_{i \leq n} x_i^*(x)x_i$ converges to y in $L_p[0,2\pi]$. But because $\{x_i\}$ is a basis for $L_p[0,2\pi]$ we have $x = y$ so that x is in $L_2[0,2\pi]$. However, this result is impossible because for instance $x(t) = t^{-\frac{1}{2}}, 0 \leq t \leq 2\pi$ is in $L_p[0,2\pi]$ but not in $L_2[0,2\pi]$. This shows the existence of an element x in $L_p[0,2\pi]$ such that $\sum_{i=1}^{\infty} |x_i^*(x)|^2 = \infty$.

Let us now suppose that $\{x_i\}$ forms an unconditional basis for $L_p[0,2\pi]$. Then there exists for each x in $L_p[0,2\pi]$ (cf. section 4) a constant $M > 0$ such that $\left\| \sum_{i \in \mu} x_i^*(x)x_i \right\| < M$ for every finite set μ of positive integers. Taking $\rho_i = \pi^{-\frac{1}{2}}(|x_{2i-1}^*(x)|^2 + |x_{2i}^*(x)|^2)^{\frac{1}{2}}$, we have for $i = 1,2,\dots$, by a suitable choice of φ_i,

$$x_{2i-1}^*(x) x_{2i-1}(t) + x_{2i}^*(x) x_{2i}(t) = \rho_i \cos(it + \varphi_i).$$

With this notation one gets

$$\int\limits_0^{2\pi} \Big|\sum_{i\in\mu} \rho_i \cos(it+\varphi_i)\Big|^p dt < M^p.$$

Let $\{\psi_n\}$ be Rademacher's orthonormal system in $L_2[0,1]$. Next, let $s\in[0,1]$ be fixed, let m be any positive integer and let μ^+ and μ^- be the set of integers i, $i\leqslant m$, for which $\psi_i(s)$ is positive, respectively negative. Then

$$\int\limits_0^{2\pi} \Big|\sum_{i\leqslant m} \rho_i \cos(it+\varphi_i)\psi_i(s)\Big|^p dt = \int\limits_0^{2\pi} \Big|\sum_{i\in\mu^+} \rho_i \cos(it+\varphi_i) - \sum_{i\in\mu^-} \rho_i \cos(it+\varphi_i)\Big|^p dt$$

$$\leqslant \left\{\sum_{\pm}\left[\int\limits_0^{2\pi}\Big|\sum_{i\in\mu^\pm} \rho_i \cos(it+\varphi_i)\Big|^p dt\right]^{1/p}\right\}^p$$

$$\leqslant (2M)^p < 4M^p.$$

As a result of an inequality for lacunary series (KACZMARZ and STEINHAUS [1], p. 132),

$$\Big(\sum_{i\leqslant m} |\alpha_i|^2\Big)^{p/2} \leqslant K_p \int\limits_0^1 \Big|\sum_{i\leqslant m} \alpha_i \Psi_i(s)\Big|^p ds,$$

where $\{\alpha_i\}$ is any set of real numbers and K_p is a positive constant depending on p only, one has

$$\Big[\sum_{i\leqslant m} \rho_i^2 \cos^2(it+\varphi_i)\Big]^{p/2} \leqslant K_p \int\limits_0^1 \Big|\sum_{i\leqslant m} \rho_i \cos(it+\varphi_i) \Psi_i(s)\Big|^p ds.$$

Integration of both sides over $(0,2\pi)$ and inverting the order of integration becomes

$$\int\limits_0^{2\pi} \Big[\sum_{i\leqslant m} \rho_i^2 \cos^2(it+\varphi_i)\Big]^{p/2} dt \leqslant 4K_p M^p.$$

Therefore, the sum in the square brackets converges with m almost everywhere in $(0,2\pi)$ to a $p/2$-th power integrable function (I.4.10). From the definition of a Lebesgue integral it follows that there exists in $(0,2\pi)$ a set E of measure $|E|=\pi$ and a constant $N>0$ such that

$$\Big[\sum_{i=1}^\infty \rho_i^2 \cos^2(it+\varphi_i)\Big]^{p/2} \leqslant N \quad \text{for every } t \text{ in } E. \text{ Integrating the sum over}$$

the set E becomes

$$\sum_{i=1}^{\infty} \rho_i^2 \int_E \cos^2(it+\varphi_i)dt = \int_E \sum_{i=1}^{\infty} \rho_i^2 \cos^2(it+\varphi_i)dt \leqslant \pi N^{2/p}$$

where we have used Lebesgue's dominated convergence theorem to justify the interchange of limit and integral. But for $i \geqslant 1$,

$$\int_E \cos^2(it+\varphi_i)dt \geqslant 4i \int_0^{\pi/(4i)} \sin^2(it)dt = 4 \int_0^{\pi/4} \sin^2 t\, dt > 0.$$

Thus $\sum_{i=1}^{\infty} \rho_i^2 < \infty$ and this contradiction shows that for $1<p<2$ the basis is not unconditional.

Finally, to treat the case $2<p<\infty$ we assume $\{x_i\}$ to be an unconditional basis for $L_p[0,2\pi]$. Then by Theorem 4.6, $\{x_i^*\}$ is an unconditional basis for $L_p^*[0,2\pi]$, so that by (I.4.6), $\{x_i\}$ is an unconditional basis for $L_q[0,2\pi]$, $1<q<2$. Since this is a contradiction the assertion of the theorem follows. Finally, the complex case results from the fact that our biorthogonal system $\{x_i,x_i^*\}$ is an (unconditional) basis for real $L_p[0,2\pi]$ if and only if it is an (unconditional) basis for complex $L_p[0,2\pi]$.

Remark. The trigonometrical system of the preceding theorem does not form a basis for $L_1[0,2\pi]$. For, supposing the contrary, the sequence of expansion operators $\{U_n\}$ of $\{x_i\}$ would be bounded (III.2.3). Since

$$(U_{2n}x)(s)=(2\pi)^{-1}\int_0^{2\pi}(\sin\tfrac{1}{2}t)^{-1}\sin(n+\tfrac{1}{2})t\,x(s-t)dt,\ 0\leqslant s\leqslant 2\pi,\ x\in L_1[0,2\pi]$$

(the formula is known as Dirichlet's integral (DUNFORD and SCHWARTZ [1], p. 359)), this would imply that $\infty > \sup_n \|U_n\| \geqslant \sup_n \sup \{\|U_{2n}x\| \,|\, \|x\|$ $\leqslant 1, x\in L_1[0,2\pi]\} \geqslant \sup_n (2\pi)^{-1}\int_0^{2\pi}(\sin\tfrac{1}{2}t)^{-1}|\sin(n+\tfrac{1}{2})t|dt$ (hint: take $x(t)=1/\varepsilon,\ 0\leqslant t\leqslant\varepsilon,\ x(t)=0$ else, and choose $\varepsilon>0$ arbitrarily small). But $\int_0^{2\pi}(\sin\tfrac{1}{2}t)^{-1}|\sin(n+\tfrac{1}{2})t|dt \geqslant 2\int_0^{\pi}s^{-1}\sin(2n+1)s\,ds \geqslant 2[1+\tfrac{1}{3}+\tfrac{1}{5}+\cdots +1/(4n+1)]$ which contradicts the above estimate. Hence $\{x_i\}$ cannot be a basis for $L_1[0,2\pi]$.

Similarly, the trigonometrical system $\{x_i\}$ cannot be a basis for $C[0,2\pi]$ either. For $(2\pi)^{-1}\int_0^{2\pi}(\sin\tfrac{1}{2}t)^{-1}|\sin(n+\tfrac{1}{2})t|dt\leqslant \sup\{|(U_{2n}x)(0)|\,|\,\|x\|\leqslant 1,$ $x\in C[0,2\pi]\}\leqslant\sup\{\|U_{2n}x\|\,|\,\|x\|\leqslant 1, x\in C[0,2\pi]\}$. The first inequality is obtained by choosing for x continuous functions of norm one in $C[0,2\pi]$ which approximate the function $\text{sign}\sin(n+\tfrac{1}{2})t$ with respect to the norm in $L_1[0,2\pi]$.

The two results can be obtained as well from the knowledge of an element x in $L_1[0,2\pi]$ whose Fourier series diverges in $L_1[0,2\pi]$

(DUNFORD and SCHWARTZ [1], p. 359), and from the existence of a function x in $C[0,2\pi]$ whose Fourier series diverges at a point of the interval $[0,2\pi]$ (ZYGMUND [1], p. 167), using the fact that $\sum\limits_{i=0}^{\infty} x_i^*(x)x_i$ is just the formal Fourier series of x.

References for Chapter III: BANACH [1], DAY [2], GELBAUM [1], JAMES [4], KARLIN [2], RETHERFORD [4], SINGER [3, 7 and 12], WILANSKY [1] and ZYGMUND [1].

CHAPTER IV

Orthogonality, Projections and Equivalent Bases

If there is a basis for a closed linear subspace Y of Banach space X, a very general condition allows to define a projection of X on Y. On the other hand, projections are very useful tools for existence proofs of bases. Indeed, they can be applied in the proof of the existence theorem of NIKOL'SKIĬ given in the first paragraph. The theorem is quite essential in the theory and in the application of the theory of bases and, accordingly, we have to make reference to it in many of the subsequent theorems and corollaries. The second paragraph refers to the very nice fact that the concept of orthogonality can be extended from Hilbert spaces to the normed linear spaces. It gives the relation between total orthogonal systems, simple $\mathcal{N}_1$-spaces and monotone bases. The last section is concerned with equivalent bases, block bases and the theorem that every infinite dimensional Banach space contains an infinite dimensional subspace with a basis.

1. Bases and Projections

Let $\{x_i, x_i^*\}$ be a basis for the Banach space X. We recall that the expansion operators U_n are defined by $U_n x = \sum_{i \leq n} x_i^*(x) x_i$ for all n and x in X (III.2.1). For each m and n, $U_m U_n = U_{\min\{m,n\}}$, implying that U_n is a projection in X. In the whole chapter IV, the word projection will be used in the sense of a continuous projection.

Theorem 1. *Let Y be a closed linear subspace of X. If $\{y_i\}$ is a basis for Y and if there is a sequence $\{x_i^*\}$ in X^* such that $x_i^*(y_j) = \delta_{ij}$ and such that $\sum_{i \leq n} x_i^*(x) y_i$ converges in X for all x in X, then P, defined by $Px = \lim_n \sum_{i \leq n} x_i^*(x) y_i$, $x \in X$, defines a projection of X on Y.*

Proof. Since the series for Px converges in X for $x \in X$ we have $\|Px\| \leq \sup_n \left\| \sum_{i \leq n} x_i^*(x) y_i \right\| < \infty$, $x \in X$. By (I.3.14) there is then a constant

M such that $\|Px\| \leqslant M\|x\|$ so that P is an endomorphism of X. Clearly, $P(X) \subset Y$. Since $\{y_i\}$ is a basis for Y there is a unique sequence $\{\alpha_i\}$ in Φ such that $y = \lim\limits_{n} \sum\limits_{i \leqslant n} \alpha_i y_i$, $y \in Y$. Due to $x_i^*(y_j) = \delta_{ij}$, $Py = \lim\limits_{n} \sum\limits_{i \leqslant n} \alpha_i P y_i$

$= \lim\limits_{n} \sum\limits_{i \leqslant n} \alpha_i y_i = y$ so that $P(X) = Y$. Thus P is a projection of X on Y, concluding the proof of the theorem.

Corollary 2. *Let* $\{x_i, x_i^*\}$ *be a basis for* X. *If* $\sum\limits_{i \leqslant n} x^{**}(x_i^*) x_i$ *converges in* X *for all* x^{**} *in* X^{**}, *then there exists a projection of* X^{**} *on* $J(X)$.

Proof. Let J' be the natural embedding of X^* into X^{***}. We take $Y = J(X) \subset X^{**}$. Then $x = \lim\limits_{n} \sum\limits_{i \leqslant n} x_i^*(x) x_i$ implies that $Jx = \lim\limits_{n} \sum\limits_{i \leqslant n} Jx(x_i^*) Jx_i$

$= \lim\limits_{n} \sum\limits_{i \leqslant n} J'x_i^*(Jx) Jx_i$, and $\{Jx_i\}$ is a basis for Y. Moreover, $\{J'x_i^*\}$ is a sequence in X^{***} such that $J'x_i^*(Jx_j) = Jx_j(x_i^*) = x_i^*(x_j) = \delta_{ij}$ and such that $\sum\limits_{i \leqslant n} J'x_i^*(x^{**}) Jx_i \left(= J \sum\limits_{i \leqslant n} x^{**}(x_i^*) x_i \right)$ converges for all x^{**} in X^{**}. Thus the assumptions of the preceding theorem are fulfilled and $J \left[\lim\limits_{n} \sum\limits_{i \leqslant n} x^{**}(x_i) x_i \right]$ defines a projection of X^{**} on $J(X)$.

Theorem 3. *Let* Y *be a closed linear subspace of* X. *If* $\{y_i\}$ *is a basis for* Y *and if* P *is a projection of* X *on* Y, *then there exists a unique sequence* $\{z_i^*\}$ *in* X^* *such that* $z_i^*(y_i) = \delta_{ij}$ *and such that* $Px = \lim\limits_{n} \sum\limits_{i \leqslant n} z_i^*(x) y_i$, $x \in X$.

Proof. Let $\{y_i^*\}$ be the associated biorthogonal sequence to $\{y_i\}$. For all i the Hahn-Banach theorem insures the existence of an extension $y_i^{*'}$ of y_i^* to the whole space X. Since $y_i^{*'} \in X^*$ we can define $P^* y_i^{*'} = z_i^* \in X^*$ and we obtain the relations $z_i^*(y_j) = P^* y_i^{*'}(y_j)$ $= y_i^{*'}(P y_j) = y_i^*(y_j) = \delta_{ij}$. Now, $Px = \lim\limits_{n} \sum\limits_{i \leqslant n} y_i^*(Px) y_i = \lim\limits_{n} \sum\limits_{i \leqslant n} y_i^{*'}(Px) y_j$

$= \lim\limits_{n} \sum\limits_{i \leqslant n} P^* y_i^{*'}(x) y_i = \lim\limits_{n} \sum\limits_{i \leqslant n} z_i^*(x) y_i$ for every x in X. It remains to show that the sequence $\{z_i^*\}$ is unique. If $\{z_i^*\}$ is another sequence with the properties of $\{z_i^*\}$, then we have $\lim\limits_{n} \sum\limits_{i \leqslant n} (z_i^* - z_i^{*'})(x) y_i = 0$ for all x in X. Multiplying through by y_j^* we get $(z_j^* - z_j^{*'})(x) = 0$ for every x in X and all j. Hence the sequence $\{z_i^*\}$ is unique and the theorem is proved.

Theorem 4. *Suppose that* $\{x_i, x_i^*\}$ *is a basis for* X *and that* P *is a projection of norm one of* X^{**} *on* $J(X)$. *If for every* x^* *in* X^* *there is a unique* x^{***} *in* X^{***} *for which* $\|x^{***}\| = \|x^*\|$ *and such that* $x^*(x)$ $= x^{***}(Jx)$ *for all* x *in* X, *then* $\sum\limits_{i \leqslant n} x^{**}(x_i^*) x_i$ *converges in* X *for all* x^{**} *in* X^{**}.

Proof. Since $\{x_i\}$ is a basis for X we have for each x^{**} in X^{**}, $J^{-1}Px^{**}$
$=\lim_n \sum_{i\leqslant n} x_i^*(J^{-1}Px^{**})x_i = \lim_n \sum_{i\leqslant n} Px^{**}(x_i^*)x_i = \lim_n \sum_{i\leqslant n} J'x_i^*(Px^{**})x_i$
$=\lim_n \sum_{i\leqslant n} P^*J'x_i^*(x^{**})x_i$, where J' is the natural embedding of X^*
into X^{***}. It remains to insure the relations $P^*J'x_i^* = J'x_i^*$, hence
that $\sum_{i\leqslant n} x^{**}(x_i^*)x_i$ converges in X for each x^{**} in X^{**}:

Let F_i be the restrictions of $J'x_i^*$ to $J(X)$. Then $\|F_i\| = \sup\{|F_i(Jx)|$
$|x\in X, \|x\| \leqslant 1\} = \sup\{|J'x_i^*(Jx)| \, |x\in X, \|x\| \leqslant 1\} = \sup\{|x_i^*(x)| \, |x\in X, \|x\| \leqslant 1\}$
$=\|x_i^*\| = \|J'x_i^*\|$. Moreover, $P^*J'x_i^*$ is an element of X^{***} and
$P^*J'x_i^*(Jx) = J'x_i^*(PJx) = J'x_i^*(Jx) = F_i(Jx)$, $x\in X$. Hence $P^*J'x_i$ is
an extension of F_i to the domain X^{**} and we have $\|F_i\| \leqslant \|P^*J'x_i^*\|$.
On the other hand, since $\|P^*\| = \|P\| = 1$. we obtain $\|P^*J'x_i^*\| \leqslant \|J'x_i^*\|$.
Thus $\|P^*J'x_i^*\| = \|J'x_i^*\| = \|x_i^*\|$ and since $x_i^*(x) = Jx(x_i^*) = J'x_i^*(Jx)$
$=P^*J'x_i^*(Jx)$ for all x in X it follows from the hypothesis that
$P^*J'x_i^* = J'x_i^*$, concluding the proof of the theorem.

Theorem 5. (NIKOL'SKIĬ) *A total sequence $\{x_i\}$ of non-zero elements
in X is a basis for X if and only if there is a constant $M \geqslant 1$ such that
$\left\|\sum_{i\leqslant n} \alpha_i x_i\right\| \leqslant M \left\|\sum_{i\leqslant m} \alpha_i x_i\right\|$ for each n,m with $n\leqslant m$ and arbitrary coefficients $\alpha_1, ..., \alpha_m$ in Φ.*

Proof. If $\{x_i\}$ is a basis for X we define the constant $M \geqslant 1$ by
$M = \sup_n \|U_n\|$ (III.2.3). Let $\{\alpha_i\}$ be any sequence in Φ and let n and m be
positive integers such that $n \leqslant m$. Then $\sum_{i\leqslant n} \alpha_i x_i = U_n \sum_{i\leqslant m} \alpha_i x_i$ and
$\left\|\sum_{i\leqslant n} \alpha_i x_i\right\| \leqslant M \left\|\sum_{i\leqslant m} \alpha_i x_i\right\|$.

Conversely, we assume the inequality of the theorem to be satisfied.
From this inequality it follows that every finite collection $\{x_i | i \leqslant n\}$
is linearly independent. Then we define the subspaces $E_1 \subset E_2 \subset ... \subset E_n \subset ...$
of X by $E_n = \mathrm{sp}\{x_i | i \leqslant n\}$. Let now P_{nn} be the identity mapping from E_n
into X. Moreover, we define the projections P_{nm} of E_m on E_n, $m > n$, by
$P_{nm}(x+y) = x$ for each x in E_n and y in $\mathrm{sp}\{x_i | n < i \leqslant m\}$. From the obvious
equation $\|P_{nn}\| = 1$ and the inequality it follows that $\|P_{nm}(x+y)\| = \|x\|$
$\leqslant M\|x+y\|$, hence that $\|P_{nm}\| \leqslant M$ for every $m \geqslant n$. Now, we define
the linear transformation $P_n': D \to E_n$ by $P_n'x = P_{nm}x$, $x\in E_m$, $m < \infty$,
where $D = \{x | x\in E_m, m < \infty\}$. Thus, $\|P_n'\| \leqslant M$ and since the sequence
$\{x_i\}$ is total in X we have $\bar{D} = X$. Hence P_n' has a unique linear extension,
say P_n, on X and $\|P_n\| \leqslant M$ (I.2.12). Since E_n is closed, P_n has range E_n
so that P_n is a projection of X on E_n.

The existence of the sequence $\{P_n\}$ enables us to show that the total
sequence $\{x_i\}$ is a basis for X: For each x in X and every $\varepsilon > 0$ we have
an index n such that $\inf\{\|x-y\| \, |y\in E_n\} < \varepsilon$. Hence there is a y in E_n for

which $\|x-y\|<\varepsilon$ and we have $\|P_m x-x\|\leqslant\|P_m x-y\|+\|y-x\|$ $=\|P_m(x-y)\|+\|x-y\|<(M+1)\varepsilon$ for all $m\geqslant n$. This implies that $P_n x$ converges strongly to x . Let $x_i^*\in X^*,\,i=1,2,\ldots$ be defined by $x_i^*(x)x_i=(P_i-P_{i-1})x,\,x\in X$, taking $P_0=0$. Since $x_i^*(x_j)x_i=(P_i-P_{i-1})x_j$ $=\delta_{ij}x_i$, $\{x_i,x_i^*\}$ is a biorthogonal system for X . Now $U_n=\sum_{i\leqslant n}(P_i-P_{i-1})=P_n$,
so that $\sup\|U_n\|\leqslant M$. Finally, since $\overline{\mathrm{sp}}\{x_i\}=X$, Corollary III.2.3 implies that $\{x_i\}$ is a basis for X and we have the theorem.

We note that the first part of Theorem III.2.2 again may easily be obtained as a corollary of the above theorem: By the principle of uniform boundedness one obtains from the hypothesis that $\sup_n\|U_n\|<\infty$. Now,
let $x=\sum_{i\leqslant m}\alpha_i x_i,\,\alpha_i\in\Phi,\,i=1,\ldots,m$. Consequently, $\left\|\sum_{i\leqslant n}\alpha_i x_i\right\|=\|U_n x\|$ $\leqslant\sup_n\|U_n\|\left\|\sum_{i\leqslant m}\alpha_i x_i\right\|$ for each $n\leqslant m$, and by the preceding theorem $\{x_i\}$ is a basis for $\overline{\mathrm{sp}}\{x_i\}$.

Theorem 6. *If* Γ *is a determining manifold for* X *and if* $\{x_n\}$ *is a sequence in* X *such that*

(i) $$0<\inf_n\|x_n\|\leqslant\sup_n\|x_n\|<\infty,$$

(ii) $$\lim_n x^*(x_n)=0\quad\text{for any }x^*\text{ in }\Gamma,$$

then there exists a subsequence $\{x_{n_i}\}$ *of* $\{x_n\}$ *with* $n_1=1$ *which is a basis for* $\overline{\mathrm{sp}}\{x_{n_i}\}$.

Proof. By (i) we obviously may assume that $\|x_n\|=1,\,n=1,2,\ldots$. Let $n_1,\ldots,n_k$ be an arbitrary increasing finite set of integers, $Z_k=\mathrm{sp}\{x_{n_i}|i\leqslant k\}$ and let $S_k=\{z\in Z_k|\,\|z\|=1\}$, the unit sphere in Z_k . Because Z_k is finite dimensional, it is clear that S_k is compact (I.1.12). Thus there are elements $z_1,\ldots,z_m$ in S_k such that $\inf\{\|z-z_i\|\,|z\in S_k,\,i\leqslant m\}<(2k)^{-2}/4$ (i.e. there exists a finite ε -net with respect to S_k). Now, there exist elements $z_1^*,\ldots,z_m^*$ in $\Gamma\subset X^*$ such that $1-(2k)^{-2}/4\leqslant|z_i^*(z_i)|\leqslant1$ and $\|z_i^*\|=1$. From the hypothesis (ii) we infer that there exists an integer $n_{k+1}>n_k$ for which $\sup\{|z_i^*(x_{n_{k+1}})|\,|i\leqslant m\}<(2k)^{-2}/4$. Next, let z be in S_k and α in $\Phi,\,|\alpha|\geqslant2$. Then $\|z+\alpha x_{n_{k+1}}\|\geqslant|\alpha|\,\|x_{n_{k+1}}\|-\|z\|\geqslant1>1-(2k)^{-2}$. On the other hand, if $|\alpha|<2$, then there is an index i for which $\|z-z_i\|$ $<(2k)^{-2}/4$, and

$$\|z+\alpha x_{n_{k+1}}\|\geqslant|z_i^*(z+\alpha x_{n_{k+1}})|\geqslant|z_i^*(z_i)|-|z_i^*(z-z_i)|-|z_i^*(\alpha x_{n_{k+1}})|$$

$$\geqslant1-(2k)^{-2}/4-\|z-z_i\|-2|z_i^*(x_{n_{k+1}})|$$

$$\geqslant1-(2k)^{-2}.$$

Starting with $n_1 = 1$ we can find in this way an increasing sequence $\{n_k\}$ such that for arbitrary α_k in Φ,

$$\left\| \sum_{i \leqslant k} \alpha_i x_{n_i} \right\| \leqslant (1 - (2k)^{-2})^{-1} \left\| \sum_{i \leqslant k+1} \alpha_i x_{n_i} \right\|,$$

for $k = 1, 2, \ldots$. Thus for $1 \leqslant p < q$ we have

$$\left\| \sum_{i \leqslant p} \alpha_i x_{n_i} \right\| \leqslant \prod_{k=p}^{q-1} (1 - (2k)^{-2})^{-1} \left\| \sum_{i \leqslant q} \alpha_i x_{n_i} \right\|.$$

Since

$$\prod_{k=p}^{q-1} (1 - (2k)^{-2})^{-1} \leqslant \prod_{k=1}^{\infty} (1 - (2k)^{-2})^{-1} = \frac{\pi/2}{\sin(\pi/2)} = \pi/2,$$

Theorem 5 implies that $\{x_{n_i}\}$ is a basis for $\overline{\mathrm{sp}}\{x_{n_i}\} \subset X$ and the theorem is verified.

Theorem 7. (GRINBLYUM)   *A sequence $\{x_i\}$ which is total in X is a basis for X if and only if there exists a positive constant α such that for all n,*

$$\mathrm{dist}(S_n, X_n) \geqslant \alpha,$$

where S_n is the unit sphere $S_n = \{x \mid x \in X_n', \|x\| = 1\}$ in $X_n' = \mathrm{sp}\{x_1, \ldots, x_n\}$, $X_n = \overline{\mathrm{sp}}\{x_{n+1}, x_{n+2}, \ldots\}$ and $\mathrm{dist}(S_n, X_n) = \inf\{\|x - y\| \mid x \in S_n, y \in X_n\}$ is the distance from S_n to X_n.

Proof. Necessity. Let $\{x_i\}$ be a basis for X which by Theorem III.2.4 and Corollary III.2.3 has a corresponding sequence $\{U_n\}$ of expansion operators, such that $1 \leqslant \sup_n \|U_n\| < \infty$. We define $\alpha \in (0,1]$ by $\alpha = \left(\sup_n \|U_n\|\right)^{-1}$. Since each U_n is a continuous projection, $(I - U_n)(X)$ is closed and since $\mathrm{sp}\{x_{n+1}, x_{n+2}, \ldots\} \subset (I - U_n)(X)$, one has $X_n \subset (I - U_n)(X)$. Moreover, it is clear that S_n is the unit sphere of $U_n(X)$. Hence

$$\mathrm{dist}(S_n, X_n) = \inf\{\|x - y\| \mid x \in U_n(X), \|x\| = 1, y \in X_n\}$$

$$\geqslant \inf\{\|x - y\| \mid x \in U_n(X), \|x\| = 1, y \in (I - U_n)(X)\}$$

$$\geqslant \alpha \inf\{\|U_n(x - y)\| \mid x \in U_n(X), \|x\| = 1, y \in (I - U_n)(X)\}$$

$$= \alpha.$$

Sufficiency. Let $\{\alpha_i\}$ be an arbitrary sequence in Φ and let n, m be any integers with $n < m$. Supposing that $\mathrm{dist}(S_n, X_n) \geqslant \alpha > 0$, it then follows that

$$\left\|\sum_{i\leqslant m}\alpha_i x_i\right\| = \left\|\left\|\sum_{i\leqslant n}\alpha_i x_i\right\|^{-1}\sum_{i\leqslant n}\alpha_i x_i + \left\|\sum_{i\leqslant n}\alpha_i x_i\right\|^{-1}\sum_{i=n+1}^{m}\alpha_i x_i\right\| \cdot \left\|\sum_{i\leqslant n}\alpha_i x_i\right\|$$

$$\geqslant \alpha\left\|\sum_{i\leqslant n}\alpha_i x_i\right\|$$

(if $\sum_{i\leqslant n}\alpha_i x_i = 0$, $\left\|\sum_{i\leqslant m}\alpha_i x_i\right\|\geqslant\alpha\left\|\sum_{i\leqslant n}\alpha_i x_i\right\|$ is trivially satisfied). Thus Theorem 5 applies with $M = \alpha^{-1} + 1$ and we are done.

2. Orthogonality, simple $\mathcal{N}_1$-Spaces and Monotone Bases

It is very useful to extend the concept of orthogonality, which is well known in Hilbert spaces, to normed linear spaces. There are several possibilities to do this and we restrict ourself to the one which seems to be the simplest and best suited for application to the theory of bases in a Banach space X:

Definition 1. *We say that an element x of X is orthogonal to an element y of X if and only if $\|x\|\leqslant\|x+\alpha y\|$ for every α in Φ.*

Clearly this definition, introduced by JAMES [3] is homogeneous in x and y but not necessarily symmetric nor additive from the left (or from the right) as it is in a Hilbert space. We have the following result for Hilbert spaces H (we use the word orthogonal throughout in the sense of Definition 1):

Theorem 2. *Two elements x and y of H are orthogonal if and only if $(x,y)=0$.*

Proof. Let $(x,y)=0$. Then $\|x\|^2 = (x+\alpha y-\alpha y, x+\alpha y-\alpha y) = \|x+\alpha y\|^2 - \|\alpha y\|^2 \leqslant \|x+\alpha y\|^2$ for every α in Φ. Conversely, let $\|x\|^2 \leqslant \|x+\alpha y\|^2$. This implies $\|\alpha y\|^2 + 2\operatorname{Re}(x,\alpha y)$ to be non-negative. But this expression has minimum value $-|(x,y)|^2/\|y\|^2$ so that (x,y) must vanish.

Definition 3. *A sequence $\{x_i\}$ in X is said to be an orthogonal system if $\operatorname{sp}\{x_i|i\leqslant n\}$ is orthogonal to x_{n+1} for every n.*

It follows that the elements $x_1, ..., x_n$ of an orthogonal system are linearly independent.

Definition 4. *A Banach space X is an $\mathcal{N}_1$-space if there is an arbitrary index set A and a family $\{N_a|a\in A\}$ of finite dimensional subspaces, directed by inclusion, whose union is dense in X and such that N_a is the range of a projection P_a of norm one of X. As a special instance, an $\mathcal{N}_1$-space is simple if A is the set of positive integers, $N_1\subset N_2\subset\cdots\subset N_n\subset\cdots$ and $\dim N_n = n$.*

Evidently, a simple $\mathcal{N}_1$-space is separable. It is known that L_p-spaces $(1 \leqslant p < \infty)$ and the space $C(S)$, where S is a compact metric space, are $\mathcal{N}_1$-spaces (MICHAEL and PELCZYNSKI [1]) and that $c_0, l_p, L_p[0,1] (1 \leqslant p < \infty)$ and separable Hilbert spaces provide examples for simple $\mathcal{N}_1$-spaces.

Theorem 5. (MAZUR) *In each simple $\mathcal{N}_1$-space X there exists a monotone basis for X.*

Proof. There exists a set $\{P_i\}$ of projections of unit norm of X such that $P_1(X) \subset P_2(X) \subset \cdots, \dim P_i(X) = i$ and $\bigcup_{i=1}^{\infty} P_i(X)$ is dense in X. Let $\{x_i\}$ be a sequence of non-zero elements in X such that $x_1 \in P_1(X)$ and $x_i \in P_i(X) \cap \operatorname{Ker} P_{i-1}$ for $i > 1$, where $\operatorname{Ker} P_i$ denotes the set $\{x \in X | P_i x = 0\}$. Moreover, let $\{\alpha_i\}$ be an arbitrary sequence in Φ. Then

$$\left\| \sum_{i \leqslant n} \alpha_i x_i \right\| = \left\| P_n \sum_{i \leqslant n+1} \alpha_i x_i \right\| \leqslant \left\| \sum_{i \leqslant n+1} \alpha_i x_i \right\|.$$

From this one gets inductively for $n \leqslant m$,

$$\left\| \sum_{i \leqslant n} \alpha_i x_i \right\| \leqslant \left\| \sum_{i \leqslant m} \alpha_i x_i \right\|$$

so that by Theorem 1.5 $\{x_i\}$ is a basis for X. $\{x_i^*\}$ is defined to be the biorthogonal sequence to $\{x_i\}$. Finally, the basis is monotone, since for $x \in X$,

$$\left\| \sum_{i \leqslant n} x_i^*(x) x_i \right\| \leqslant \lim_n \left\| \sum_{i \leqslant n} x_i^*(x) x_i \right\|$$

$$= \left\| \lim_n \sum_{i \leqslant n} x_i^*(x) x_i \right\| = \|x\|.$$

Theorem 6. *If $\{x_i\}$ is a monotone basis for X, then $\{x_i\}$ is a total orthogonal system.*

Proof. If $\{x_i, x_i^*\}$ is a basis for X then the sequence $\{x_i\}$ is obviously total in X. The monotony implies that $\left\| \sum_{i \leqslant n} x_i^*(x) x_i \right\| \leqslant \left\| \sum_{i \leqslant n+1} x_i^*(x) x_i \right\|$ for each x in X. We assume $y \in \operatorname{sp}\{x_i | i \leqslant n\}$ and $\alpha \in \Phi$, both arbitrary, and take $x = y + \alpha x_{n+1}$. Since $x = \sum_{i \leqslant n+1} x_i^*(x) x_i$ and the x_i's are linearly independent one has $y = \sum_{i \leqslant n} x_i^*(x) x_i$ and $\alpha = x_{n+1}^*(x)$. Hence $\|y\| \leqslant \|y + \alpha x_{n+1}\|$, showing the orthogonality property. This completes the proof of the theorem.

Theorem 7. *A Banach space with a total orthogonal system is a simple $\mathcal{N}_1$-space.*

Proof. Let $\{x_i\}$ be a total orthogonal system in the Banach space X and define $N_n = \mathrm{sp}\{x_i | i \leqslant n\}$. Obviously, $N_n \subset N_{n+1}$, $n = 1, 2, \ldots$ and $\dim N_n = n$. Now by Nikol'skiĭ's theorem, $\{x_i\}$ is a basis for X which, clearly, must be monotone. Moreover, it follows that $\left\| \sum_{i \leqslant n} x_i^*(x) x_i \right\| \leqslant \|x\|$

for all n and x in X, where $\{x_i^*\}$ is the associated biorthogonal sequence to $\{x_i\}$. Hence $P_n : X \rightarrow X$, given by $P_n(x) = \sum_{i \leqslant n} x_i^*(x) x_i$, $x \in X$, is a

projection of norm one of X onto N_n which implies that X is a simple $\mathcal{N}_1$-space.

As a result of the preceding three theorems we have

Corollary 8. *In a Banach space* X *the following three statements are equivalent*:
(i) X *is a simple* $\mathcal{N}_1$-*space*.
(ii) *There exists a monotone basis* $\{x_i\}$ *for* X.
(iii) $\{x_i\}$ *is a total orthogonal system in* X.

3. Equivalent Bases

Definition 1. *Let* X *and* Y *be Banach spaces,* $\{x_i\}$ *a basis for* X *and* $\{y_i\}$ *a basis for* Y. *Then* $\{x_i\}$ *and* $\{y_i\}$ *are said to be equivalent if for a sequence* $\{\alpha_i\} \subset \Phi$ *the convergence of* $\sum_{i \leqslant n} \alpha_i x_i$ *in* X *is equivalent to the convergence of* $\sum_{i \leqslant n} \alpha_i y_i$ *in* Y.

Theorem 2. *The bases* $\{x_i\}$ *for* X *and* $\{y_i\}$ *for* Y *are equivalent if and only if there is a topological isomorphism* T *of* X *onto* Y *such that* $Tx_i = y_i$, $i = 1, 2, \ldots$.

Proof. Sufficiency. Let $\sum_{i=1}^{\infty} \alpha_i x_i$ be the unique expansion for some x in X. Since T is a topological isomorphism, $Tx = \sum_{i=1}^{\infty} \alpha_i Tx_i$ and this expansion is unique, because $\sum_{i=1}^{\infty} \alpha_i Tx_i = 0$ implies $x = 0$ and hence $\alpha_i = 0$ for all i. Taking $y = Tx$ and $y_i = Tx_i$, $i = 1, 2, \ldots$, every y in Y thus has the unique expansion $y = \sum_{i=1}^{\infty} \alpha_i y_i$ and the sufficiency is verified.

Necessity. From the proof of Theorem III.1.3 we know that X is topologically isomorphic to a Banach space X' and that Y is topologically isomorphic to a Banach space Y'; X' and Y' consisting of the vector space of all sequences $\{\alpha_i\}$ in Φ for which $\lim_n \sum_{i \leqslant n} \alpha_i x_i$ and $\lim_n \sum_{i \leqslant n} \alpha_i y_i$ respectively, exist, and which have norm $\sup_n \left\| \sum_{i \leqslant n} \alpha_i x_i \right\|$ and $\sup_n \left\| \sum_{i \leqslant n} \alpha_i y_i \right\|$ respectively. The hypothesis implies that the identity mapping I' of X'

into Y' is onto. Since the vanishing of the norm of $\{\alpha_i\}$ in Y' implies $\alpha_i = 0$ for all i, hence implies the vanishing of the norm of $\{\alpha_i\}$ in X', I' is one-to-one. The following argument leads to the conclusion that I' is closed: Let $\{\alpha_i\}, \{\beta_i\}, \{\alpha_{mi}\} \in X'$, $m = 1, 2, \ldots$ be such that $\lim\sup\limits_{m} \left\| \sum\limits_{i \leq n} (\alpha_{mi} - \alpha_i) x_i \right\| = 0$ and that $\lim\sup\limits_{m} \left\| \sum\limits_{i \leq n} (\alpha_{mi} - \beta_i) y_i \right\| = 0$. Since $\{x_i\}$ and $\{y_i\}$ are bases, uniqueness shows that $x_n, y_n \neq 0$ for each fixed n. For every $\varepsilon > 0$ we then have an m for which

$$\sup_n \max \left\{ \left\| \sum_{i \leq n} (\alpha_{mi} - \alpha_i) x_i \right\|, \quad \left\| \sum_{i \leq n} (\alpha_{mi} - \beta_i) y_i \right\| \right\} < \varepsilon/2 \, .$$

Therefore,

$$|\alpha_n - \beta_n| \leq |\alpha_{mn} - \alpha_n| + |\alpha_{mn} - \beta_n| \leq \frac{1}{\|x_n\|} \|(\alpha_{mn} - \alpha_n) x_n\|$$

$$+ \frac{1}{\|y_n\|} \|(\alpha_{mn} - \beta_n) y_n\|$$

$$\leq \varepsilon/\|x_n\| + \varepsilon/\|y_n\|$$

and we have $\{\alpha_i\} = \{\beta_i\}$, i.e. I' is closed. But I' is defined on the whole of X', hence I' is bounded (I.2.10). From these properties of I' one concludes that X' and Y' are topologically isomorphic so that X is topologically isomorphic to Y (I.2.6). Finally, since the topological isomorphisms T_X and T_Y of X (of Y) onto X' (onto Y' respectively) are such that $T_X x_i = \{\delta_{ij}\}$ and that $T_Y y_i = \{\delta_{ij}\}$ it is clear that the transformation $T: X \to Y$ of the theorem can be defined by $Tx = T_Y^{-1} I' T_X x$, $x \in X$, hence such that $Tx_i = y_i$ for all i and the theorem is verified.

Remark. The theorem remains true if X and Y are supposed to be locally convex complete linear metric spaces (see ARSOVE [5], Theorem 1).

Theorem 3. *Let* $\{x_i, x_i^*\}$ *be a basis for* X. *If the sequence* $\{y_i\}$ *in* X *satisfies* $\sum\limits_{i=1}^{\infty} \|x_i^*\| \, \|x_i - y_i\| < 1$, *then* $\{y_i\}$ *is an equivalent basis for* $\overline{sp}\{y_i\}$.

Proof. For $i \leq m$ one has for arbitrary α_i's in Φ,

$$|\alpha_i| = \left| \sum_{j \leq m} x_i^*(\alpha_j x_j) \right| \leq \left\| \sum_{j \leq m} \alpha_j x_j \right\| \|x_i^*\| \, .$$

Thus

$$\left\| \sum_{i \leq m} \alpha_i y_i \right\| \leq \left\| \sum_{i \leq m} \alpha_i x_i \right\| + \sum_{i=1}^{m} |\alpha_i| \, \|x_i - y_i\|$$

$$\leq \left\| \sum_{i \leq m} \alpha_i x_i \right\| \left| 1 + \sum_{i=1}^{\infty} \|x_i^*\| \, \|x_i - y_i\| \right|$$

$$\leq 2 \left\| \sum_{i \leq m} \alpha_i x_i \right\| .$$

By Theorem 1.5 there is a constant $M \geqslant 1$ such that

$$\left\| \sum_{i \leqslant m} \alpha_i x_i \right\| \leqslant M \left\| \sum_{i \leqslant n} \alpha_i x_i \right\|$$

for any m, n and $m \leqslant n$. Hence

$$\left\| \sum_{i \leqslant m} \alpha_i y_i \right\| \leqslant 2M \left\| \sum_{i \leqslant n} \alpha_i x_i \right\|$$

$$\leqslant 2M \left(\left\| \sum_{i \leqslant n} \alpha_i y_i \right\| + \sum_{i=1}^{n} |\alpha_i| \, \|x_i - y_i\| \right)$$

$$\leqslant 2M \left\| \sum_{i \leqslant n} \alpha_i y_i \right\| + 2M \left\| \sum_{i \leqslant n} \alpha_i x_i \right\| \sum_{i=1}^{\infty} \|x_i^*\| \, \|x_i - y_i\|$$

$$\leqslant K \left\| \sum_{i \leqslant n} \alpha_i y_i \right\|,$$

where

$$K = 2M \left(1 - \sum_{i=1}^{\infty} \|x_i^*\| \, \|x_i - y_i\| \right)^{-1} < \infty.$$

Again by Theorem 1.5, this implies that $\{y_i\}$ is a basis for $\overline{\mathrm{sp}}\{y_i\}$. Since the above estimates show that

$$\left\| \sum_{i=m}^{n} \alpha_i y_i \right\| \leqslant 2 \left\| \sum_{i=m}^{n} \alpha_i x_i \right\|$$

$$\leqslant \frac{K}{M} \left\| \sum_{i=m}^{n} \alpha_i y_i \right\|,$$

it is clear that $\{x_i\}$ is equivalent to $\{y_i\}$.

Theorem 4. *Let Y and Z be closed linear subspaces of X and let $\{y_i\}$ and $\{z_i\}$ be bases for Y and Z respectively. Moreover, let P be a projection of X on Y and let $\{y_i^*\}$ be the associated biorthogonal sequence to $\{y_i\}$. If the condition $\|P\| \sum_{i=1}^{\infty} \|y_i^*\| \, \|y_i - z_i\| < 1$ is verified, then there exists a projection of X on Z.*

Proof. Let $T: X \to X$ be defined by

$$Tx = x - Px + \sum_{i=1}^{\infty} y_i^*(Px) z_i, \qquad x \in X.$$

Since $\left\|\sum_{i=m}^{n} y_i^*(Px)z_i\right\| \leqslant \left\|\sum_{i=m}^{n} y_i^*(Px)y_i\right\| + \|Px\| \sum_{i=m}^{n} \|y_i^*\| \|y_i - z_i\|$, the limit exist. From

$$\|I - T\| = \sup\left\{\left\|Px - \sum_{i=1}^{\infty} y_i^*(Px)z_i\right\| \;\middle|\; x \in X, \|x\| \leqslant 1\right\}$$

$$= \sup\left\{\left\|\sum_{i=1}^{\alpha} y_i^*(Px)(y_i - z_i)\right\| \;\middle|\; x \in X, \|x\| \leqslant 1\right\}$$

$$\leqslant \|P\| \sum_{i=1}^{\infty} \|y_i^*\| \|y_i - z_i\| < 1,$$

it follows that $0 < 1 - \|I - T\| \leqslant \|T\| \leqslant 1 + \|I - T\| < 2$. Thus T is a topological isomorphism of X onto itself (I.2.15). From the definition of T it is easy to see that $T(Y) \subset Z$. Moreover, T has the property that $\{Ty_i\} = \{z_i\}$. Because for any $z \in Z$ with $z = \sum_{i=1}^{\infty} \alpha_i z_i$, $\sum_{i \leqslant n} \alpha_i y_i = \sum_{i \leqslant n} \alpha_i T^{-1} z_i = T^{-1} \sum_{i \leqslant n} \alpha_i z_i$ converges with n to an element, say y, of Y and since then $Ty = z$, T maps Y onto Z. Finally, by $TPT^{-1}TPT^{-1} = TPT^{-1}$, it follows immediately that TPT^{-1} is the desired projection of X on Z.

Definition 5. *Let $\{x_i\}$ be a basis for X, $\{\alpha_i\}$ a sequence in Φ and $\{p_n\}$ a strictly increasing sequence of positive integers. A sequence $\{y_n\}$ of non-zero elements in X, given by $y_n = \sum_{i = p_n + 1}^{p_{n+1}} \alpha_i x_i$, is called a block basis with respect to $\{x_i\}$.*

Theorem 6. *A block basis $\{y_n\}$ with respect to a basis $\{x_i\}$ for X is a basis for $\overline{\mathrm{sp}}\{y_n\}$.*

Proof. This follows immediately from Theorem 1.5.

Theorem 7. (BESSAGA-PELCZYNSKI) *If $\{x_i, x_i^*\}$ is a basis for X and if there is a sequence $\{y_n\}$ in X such that $\inf_n \|y_n\| > 0$ and $\lim_n x_i^*(y_n) = 0$ for all i, then there exists a subsequence $\{y_{p_n}\}$ of $\{y_n\}$ which is a basis for $\overline{\mathrm{sp}}\{y_{p_n}\}$ and which is equivalent to a block basis with respect to $\{x_i\}$.*

Proof. By hypothesis there exists an $\varepsilon > 0$ with $\inf_n \|y_n\| \geqslant \varepsilon$ and strictly increasing sequences of integers $\{p_n\}$ and $\{q_n\}$ such that for $n \geqslant 1$ (starting with $q_1 = 1$),

$$8M\left\|\sum_{i=1}^{q_n} x_i^*(y_{p_n})x_i\right\| \leqslant \varepsilon/2^n$$

and

$$8M\left\|\sum_{i=q_{n+1}+1}^{\infty} x_i^*(y_{p_n})x_i\right\| \leqslant \varepsilon/2^n,$$

where $M \geqslant 1$ is the constant which occurs in Theorem 1.5. Let

$$z_n = \sum_{i=q_n+1}^{q_{n+1}} x_i^*(y_{p_n})x_i, \qquad n \geqslant 1.$$

Then we have for $n \geqslant 1$,

$$\varepsilon \leqslant \|y_{p_n}\| = \left\|\sum_{i=1}^{\infty} x_i^*(y_{p_n})x_i\right\|$$

$$\leqslant \left\|\sum_{i=1}^{q_n} x_i^*(y_{p_n})x_i\right\| + \|z_n\| + \left\|\sum_{i=q_{n+1}+1}^{\infty} x_i^*(y_{p_n})x_i\right\| \leqslant \varepsilon/8 + \|z_n\|.$$

Therefore $\inf_n \|z_n\| > \varepsilon/2$. Moreover,

$$\sum_{n=1}^{\infty} \|y_{p_n} - z_n\| \leqslant \sum_{n=1}^{\infty} \left[\left\|\sum_{i=1}^{q_n} x_i^*(y_{p_n})x_i\right\| + \left\|\sum_{i=q_{n+1}+1}^{\infty} x_i^*(y_{p_n})x_i\right\|\right]$$

$$\leqslant \frac{2\varepsilon}{8M} \sum_{n=1}^{\infty} 2^{-n} = \frac{\varepsilon}{4M}.$$

Since $\{z_n\}$ is a block basis with respect to $\{x_i\}$, there exists to $\{z_n\}$ a biorthogonal sequence $\{z_n^*\}$ in $\overline{\mathrm{sp}}\{z_n\}^*$. Now, using Theorem 1.5 (cf. also Corollary III.2.3), one has for all m,

$$\|z_m^*\| = \sup\{|z_m^*(z)| \, \|z\| \leqslant 1, z \in \overline{\mathrm{sp}}\{z_n\}\}$$

$$= 2M \sup\{|z_m^*(z)| \, |2M\|z\| \leqslant 1, z \in \overline{\mathrm{sp}}\{z_n\}\}$$

$$\leqslant 2M \sup\left\{|z_m^*(z)|\Big/\left(\left\|\sum_{i \leqslant m} z_i^*(z)z_i\right\| + \left\|\sum_{i \leqslant m-1} z_i^*(z)z_i\right\|\right)\Big/ 0 \neq z \in \overline{\mathrm{sp}}\{z_n\}\right\}$$

$$\leqslant 2M/\|z_m\| < 4M/\varepsilon.$$

This shows that

$$\sum_{n=1}^{\infty} \|z_n^*\| \, \|y_{p_n} - z_n\| < 1.$$

Hence by Theorem 3, $\{y_{p_n}\}$ is a basis for $\overline{\mathrm{sp}}\{y_{p_n}\}$ which is equivalent to the block basis $\{z_n\}$.

Corollary 8. *Let Y be an infinite dimensional closed linear subspace of X and let $\{x_i\}$ be a basis for X. Then Y contains an infinite dimensional closed linear subspace with a basis which is equivalent to a block basis with respect to $\{x_i\}$.*

Proof. Let $\{x_i^*\}$ be the biorthogonal sequence belonging to $\{x_i\}$. For each n there exists a y_n in Y of the form $y_n = \lim\limits_m \sum\limits_{i=n}^{m} \alpha_i x_i$, $\alpha_i \in \Phi$, such that $\|y_n\| = 1$. Otherwise, for some n, we would have $x_i^*(y_n) = 0$ for all $i > n$, i.e. $Y \subset \mathrm{sp}\{x_j | j \leqslant n\}$, and this is impossible since Y is infinite dimensional. Thus there is a sequence $\{y_n\}$ of unit norm in Y such that $x_i^*(y_n) = 0$ for $n > i$, hence such that $\lim\limits_n x_i^*(y_n) = 0$ for all i. To complete the proof it remains to apply Theorem 7.

The next two corollaries are consequences of the following interesting lemma which is due to Banach and Mazur.

Lemma 9. *If X is a separable Banach space, then X is isometrically isomorphic to a closed subspace of the Banach space $C[0,1]$.*

Proof. Let U^* be the unit ball of X^*. Then, by Theorem I.3.21, U^* is sequentially compact in the weak* topology of X^*. Hence there exists a continuous transformation of Cantor's triadic point set P in $[0,1]$, onto U^* (HAUSDORFF [1], p. 134 and 197). Since T is continuous on P we can extend it linearly to the whole interval $[0,1]$, the result T' being a continuous transformation on $[0,1]$ onto U^*. Therefore, $T'(x) \in C[0,1]$ for every x in X. Now, there is a point t in P such that $|T'(x)| = \|x\|$ (I.3.10), where T_t' is the functional in U^* corresponding to the point t. But since $\|T'(x)\| = \sup\{|T_t'(x)| \,|\, t \in [0,1]\} \leqslant \|x\|$ we have $\|T'(x)\| = \|x\|$. Evidently $T'': X \to C[0,1]$, defined by $T''x = T_t'(x)$, $t \in [0,1]$, is then linear and isometric (i.e. $\|T''x\| = \|x\|$ for all x in X). This shows finally (I.2.15) that T'' is an isometric isomorphism and that $T''(X) \subset C[0,1]$.

Corollary 10. *Each infinite dimensional Banach space X contains an infinite dimensional subspace with a basis.*

Proof. Without loss of generality we may assume that X is separable. By the foregoing lemma, there exists an isometrical isomorphism T of X into $C[0,1]$ and, by Theorem III.7.5 this space has a basis. From the preceding corollary we know that $T(X)$ contains an infinite dimensional linear subspace Y with a basis, say $\{y_i\}$. Since T is an isometric isomorphism, it is clear that $\{T^{-1}y_i\}$ is a basis for the infinite dimensional subspace $T^{-1}(Y)$ of X.

Corollary 11. *If a sequence $\{y_n\}$ in a separable Banach space X converges weakly to 0 and $\inf\limits_n \|y_n\| > 0$, then a subsequence $\{y_{p_n}\}$ of $\{y_n\}$ forms a basis for $\overline{\mathrm{sp}}\{y_{p_n}\}$.*

Proof. Let T be the transformation used in the proof of the above corollary and let $\{z_i^*\}$ be the associated biorthogonal sequence of a basis for $C[0,1]$. Since T maps X into $C[0,1]$, T^* maps $C^*[0,1]$ into X^*. Thus

$\lim_{n} z_i^*(Ty_n) = \lim_{n} T^* z_i^*(y_n) = 0$ for all i, for y_n converges weakly to zero. Due to $\inf_{n} \|Ty_n\| = \inf_{n} \|y_n\| > 0$ and Theorem 7 there exists a subsequence $\{y_{p_n}\}$ of $\{y_n\}$ such that $\{Ty_{p_n}\}$ is a basis for $\overline{sp}\{Ty_{p_n}\}$. This finally shows that $\{y_{p_n}\}$ is a basis for $\overline{sp}\{y_{p_n}\}$.

References for Chapter IV: BESSAGA [1], BESSAGA and PELCZYNSKI [2], DAY [3], GELBAUM [1], GRINBLYUM [1], JAMES [1–3], LINDENSTRAUSS [1], MICHAEL and PELCZYNSKI [1], NIKOL'SKIĬ [1] and SINGER [12].

CHAPTER V

Bases and Structure of the Space

From the hypothesis of the existence of a basis of a certain type for a Banach space X, conclusions can be drawn on the structure of X, e. g. properties of X like weak sequential completeness, separability, reflexivity, weak conditional compactness and the dimension (finite or infinite). In the first section some results are established on the first two properties of X or of X^*. The following paragraph has to do with reflexivity. In both sections the hypothesis of an unconditional basis plays an important role, while in the last paragraph criteria for finite dimension of X are given in terms of absolutely convergent and uniform bases for X.

1. Bases, Completeness and Separability

Theorem 1. *A Banach space X with a basis is separable.*

Proof. Evidently, a basis for X is a total sequence in X. Therefore, by Theorem I.1.11, X is separable.

Corollary 2. *If X^* possesses a weak* Schauder basis, then X is separable.*

Proof. The corollary is an immediate consequence of the above theorem and of Theorem III.2.7.

Theorem 3. *If X has a boundedly complete unconditional basis, then X is weakly sequentially complete.*

Proof. Let $\{x_i, x_i^*\}$ be the basis for X and let $\{y_i\}$ be a weakly convergent sequence in X (which must not necessarily converge in the weak topology of X). Then there is a constant K such that $\sup_j \|y_j\| \leqslant K$ (I.3.15). Next, we define $\alpha_i = \lim_j x_i^*(y_j)$ for all i. From this it follows that

$$\lim_j \left\| \sum_{i \leqslant n} (\alpha_i - x_i^*(y_j)) x_i \right\| = 0$$

for all n. Thus $\left\| \sum\limits_{i \leqslant n} \alpha_i x_i \right\| \leqslant \sup\limits_j \left\| \sum\limits_{i \leqslant n} x_i^*(y_j) x_i \right\| \leqslant M \sup\limits_j \|y_j\| \leqslant M K$, where $M = \sup\limits_n \|U_n\|$ and U_n are the expansion operators of the basis $\{x_i\}$ for X. Since $\{x_i\}$ is boundedly complete, $\sum\limits_{i \leqslant n} \alpha_i x_i$ converges to an element x in X and one has $\alpha_i = x_i^*(x)$.

Now, if y_j does not converge weakly to x, there is an x^* in X^* of norm one and for some $\varepsilon > 0$ there is a subsequence $\{y_j'\}$ of $\{y_j\}$ such that with $z_j = x - y_j'$, $\mathrm{Re}\, x^*(z_j) > \varepsilon$ for all j. Again since $\{x_i\}$ is a basis for X, and due to hypothesis, there are for every $\varepsilon' > 0$ increasing sequences $\{n_j\}$ and $\{m_j\}$ such that $\|z_{n_j} - U_{m_j} z_{n_j}\| < \varepsilon'$ and $\|U_{m_j} z_{n_{j+1}}\| < \varepsilon'$. Furthermore, let $\{\beta_j\}$ be any sequence in Φ. Using Lemmas III.4.2 and 3, we obtain with a constant C, $0 < C \leqslant 1$, taking $U_{m_0} = 0$,

$$\left\| \sum_{j=1}^{n} \beta_j z_{n_j} \right\| \geqslant \left\| \sum_{j=1}^{n} \beta_j (U_{m_j} - U_{m_{j-1}}) z_{n_j} \right\| - \left\| \sum_{j=1}^{n} \beta_j (z_{n_j} - U_{m_j} z_{n_j}) \right\|$$

$$- \left\| \sum_{j=1}^{n} \beta_j U_{m_{j-1}} z_{n_j} \right\|$$

$$> C \left\| \sum_{j=1}^{n} |\beta_j| (U_{m_j} - U_{m_{j-1}}) z_{n_j} \right\| - 2\varepsilon' \sum_{j=1}^{n} |\beta_j|$$

$$\geqslant C \sum_{j=1}^{n} |\beta_j| \mathrm{Re}\, x^* [(U_{m_j} - U_{m_{j-1}}) z_{n_j}] - 2\varepsilon' \sum_{j=1}^{n} |\beta_j|$$

$$\geqslant C \sum_{j=1}^{n} |\beta_j| [\mathrm{Re}\, x^*(z_{n_j}) - \|z_{n_j} - U_{m_j} z_{n_j}\| - \|U_{m_{j-1}} z_{n_j}\|]$$

$$- 2\varepsilon' \sum_{j=1}^{n} |\beta_j|$$

$$> [C\varepsilon - 2(C+1)\varepsilon'] \sum_{j=1}^{n} |\beta_j|.$$

But

$$\left\| \sum_{j=1}^{n} \beta_j z_{n_j} \right\| \leqslant \sum_{j=1}^{n} |\beta_j| \sup_i \|z_{n_i}\| \leqslant (\|x\| + K) \sum_{j=1}^{n} |\beta_j|.$$

We can take $\varepsilon' < \varepsilon C/2(C+1)$ so that the linear transformation $T : l_1 \to X$ (naturally, l_1 is over the same field as X), defined by $T\{\gamma_i\} = \lim\limits_n \sum\limits_{i \leqslant n} \gamma_i z_{n_i}$, $\{\gamma_i\} \in l_1$, is bounded. Since $\|T\{\gamma_i\}\| \geqslant [C\varepsilon - 2(C+1)\varepsilon'] \|\{\gamma_i\}\|$, $\{\gamma_i\} \in l_1$, T has a bounded inverse (I.2.15). Hence T is a topological isomorphism of l_1 with $T(l_1)$. $T(l_1)$ is then weakly sequentially complete, because l_1 is (I.4.b, cf. also proof of Corollary V.3.1). Since by the Hahn-Banach theorem every f in $T(l_1)^*$ has a bounded linear extension to X, since $\{z_{n_j}\} \subset T(l_1)$ (due to $z_{n_j} = T\{\delta_{ji}\}$), and since by hypothesis $\lim_j x^*(z_{n_j})$ exists for all x^* in X^*, $\lim_j f(z_{n_j})$ exists for all f in $T(l_1)^*$. Hence there is a z in $T(l_1)$ for which $\lim_j x^*(z_{n_j}) = x^*(z)$, $x^* \in X^*$. But then one has

$$z = \lim_n \sum_{i \leqslant n} x_i^*(z) x_i = \lim_n \sum_{i \leqslant n} x_i \lim_j x_i^*(z_{n_j}) = \lim_n \sum_{i \leqslant n} x_i \left[x_i^*(x) - \lim_j x_i^*(y_{n_j}') \right]$$
$$= \lim_n \sum_{i \leqslant n} x_i \left[x_i^*(x) - \lim_j x_i^*(y_j) \right] = \lim_n \sum_{i \leqslant n} x_i \left[\alpha_i - \lim_j x_i^*(y_j) \right] = 0. \text{ Conse-}$$

quently, contrary to our assumption, y_n', and hence y_j, must converge weakly to x. Thus X is weakly sequentially complete and we are done.

Applying Corollary III.4.7 we immediately obtain

Corollary 4. *If X possesses an unconditional basis and if X^* is separable, then X^* is weakly sequentially complete.*

A further corollary one gets, from the combination of the above with Theorem III.4.8.

Corollary 5. *If $\{x_i, x_i^*\}$ is an unconditional basis for X, then $\{x_i^*\}$ is an unconditional basis for X^* if and only if X^* is weakly sequentially complete.*

Theorem 6. *If X has an unconditional basis, then X is weakly sequentially complete if and only if no subspace of X is topologically isomorphic with c_0.*

Proof. The necessity is proved just in the first part of the proof for Theorem III.4.5. The sufficiency is clear by Theorem 3 if the condition that no subspace of X is topologically isomorphic with c_0 implies that each unconditional basis for X is boundedly complete. But this implication is proved exactly in the second part of the proof for Theorem III.4.5 and this concludes the proof of our theorem.

Theorem 7. *If X has an unconditional basis, then the basis is shrinking if and only if no subspace of X is topologically isomorphic to l_1.*

Proof. Sufficiency. We need only to show that if the basis is not shrinking, then there is a subspace of X which is topologically isomorphic to l_1. But this is established exactly in the principal part of the proof for Theorem III.4.4.

Necessity. We assume that the basis is shrinking. Then Theorem III.3.4. warrants the existence of a basis for X^* and, evidently, X^* is separable. Now, if a subspace of X is topologically isomorphic to l_1, then by (I.3.8 and 25), a factor space of X^* is topologically isomorphic to the non-separable space l_1^* (cf. I.4.c). By this contradiction no subspace of X can be topologically isomorphic to l_1 and the theorem now follows.

2. Bases and Reflexivity

Theorem 1. *Let $\{x_i, x_i^*\}$ be a basis for X such that $\overline{\mathrm{sp}}\,\{x_i^*\} = X^*$. Then X is reflexive if and only if the basis is boundedly complete.*

Proof. By Corollary III.3.6 it is clear that $\{x_i, x_i^*\}$ is a shrinking basis for X and according to Corollary III.3.5, $\{x_i^*, J x_i\}$ then is a basis for X^*. Now let X be reflexive. Since $\overline{\mathrm{sp}}\{x_i\} = X$ this implies that $\overline{\mathrm{sp}}\,\{J x_i\} = X^{**}$ so that $\{x_i^*, J x_i\}$ is also shrinking. Thus, by Theorem III.3.9, $\{J x_i\}$ is a boundedly complete basis for X^{**}. Since X is reflexive and J is an isometric isomorphism of X onto X^{**}, $\{x_i\}$ is a boundedly complete basis for X.

Conversely, let $\{x_i, x_i^*\}$ be boundedly complete and let U and U^{**} be the unit balls in X and X^{**} respectively. Because $J(U)$ is weak* dense in U^{**} (I.3.22), it follows that for every x^{**} in U^{**} there is a sequence $\{y_n\}$ in U such that $x^{**}(x^*) = \lim_n J y_n(x^*)$ for each x^* in X^*. Thus $\sum_{i \leq j} x^{**}(x_i^*) x_i = \lim_n \sum_{i \leq j} J y_n(x_i^*) x_i = \lim_n \sum_{i \leq j} x_i^*(y_n) x_i = \lim_n U_j y_n$ and by Corollary III.2.3, $\left\| \sum_{i \leq j} x^{**}(x_i^*) x_i \right\| \leq \sup_n \|U_n\| < \infty$. But the bounded completeness of the basis then implies the existence of an x in X such that $x = \lim_n \sum_{i \leq n} x^{**}(x_i^*) x_i$. Since $(x^{**} - J x)(x_j^*) = x^{**}(x_j^*)$ $- \lim_n \sum_{i \leq n} x^{**}(x_i^*) J x_i(x_j^*) = x^{**}(x_j^*) - \lim_n \sum_{i \leq n} x^{**}(x_i^*) x_j^*(x_i) = 0$ for all x_j^* and since $\overline{\mathrm{sp}}\,\{x_i^*\} = X^*$. we have $x^{**} \in J(X)$ for every x^{**} in X^{**}. Hence X is reflexive as we wished to prove.

Theorem 2. (JAMES) *A Banach space X having a basis is reflexive if and only if the basis is both shrinking and boundedly complete.*

Proof. Let $\{x_i, x_i^*\}$ be a shrinking basis for X. Then, by Corollary III.3.6, $\overline{\mathrm{sp}}\,\{x_i^*\} = X^*$. If the basis is also boundedly complete, by the last theorem, X must be reflexive. On the other hand, suppose that X is reflexive and has a basis $\{x_i, x_i^*\}$. From Corollary III.2.9 then follows that $\{x_i^*, J x_i\}$ is a basis for X^*. On account of Theorem III.3.4, $\{x_i\}$ is shrinking. Clearly $\overline{\mathrm{sp}}\,\{x_i^*\} = X^*$. Hence Theorem 1 implies that $\{x_i, x_i^*\}$ is boundedly complete and the theorem is proved.

Theorem 3. (GELBAUM) *If X has a basis $\{x_i, x_i^*\}$, X^* is weakly sequentially complete and $\sum_{i \leq n} x^{**}(x_i^*) x_i$ converges in X for all x^{**} in X^{**}, then X is reflexive.*

Proof. We take an arbitrary $y^* \in X^*$ and define the sequence $\{y_n^*\}$ by $y_n^* = \sum_{i \leq n} y^*(x_i) x_i^*$. By hypothesis

$$x^{**}(y_n^*) = \sum_{i \leqslant n} y^*(x_i) x^{**}(x_i^*) = y^* \left(\sum_{i \leqslant n} x^{**}(x_i^*) x_i \right)$$

converges for each x^{**} in X^{**} to $x^{**}(z^*)$, where z^* is an element in X^*. But one has

$$z^*(x) = J x(z^*) = \lim_n J x(y_n^*) = \lim_n y^* \left(\sum_{i \leqslant n} x_i^*(x) x_i \right) = y^*(x)$$

for all x in X, implying $y^* = z^*$. Hence $x^{**}(y^*) = \lim_n x^{**} \left(\sum_{i \leqslant n} J x_i(y^*) x_i^* \right)$

for each x^{**} in X^{**} and since $J x_j(x_i^*) = x_i^*(x_j) = \delta_{ij}$, $\{x_i^*, J x\}$ is a weak basis for X^* which, by Corollary III.2.5, is a basis for X^*. But $\{x_i^*, J x_i\}$ is a retro-basis for X^* and we see, using Theorem III.2.12 that X is reflexive.

Theorem 4. *If X^* has a retro-basis $\{x_i^*\}$ with associated biorthogonal sequence $\{J x_i\}$ and if $\sum_{i \leqslant n} x^{**}(x_i^*) x_i$ converges in X for all x^{**} in X^{**}, then X is reflexive.*

The theorem is a corollary to Theorem III.2.12.

Let J' be the natural embedding of X^* into X^{***}, defined in the same manner as the natural embedding $J : X \to X^{**}$

Lemma 5. *There is a projection of norm one of X^{***} on $J'(X^*)$.*

Proof. We use the linear transformation $Q : X^{***} \to X^*$, given by the condition $Q x^{***}(x) = x^{***}(J x)$, $x^{***} \in X^{***}$, $x \in X$. Evidently, $\|Q x^{***}\|$ $= \sup \{|x^{***}(J x)| \,|\, x \in X, \|x\| \leqslant 1\} \leqslant \sup \{|x^{***}(x^{**})| \,|\, x^{**} \in X^{**}, \|x^{**}\| \leqslant 1\}$ $= \|x^{***}\|$. Since $Q J' x^*(x) = J' x^*(J x) = J x(x^*) = x^*(x)$, $x^* \in X^*$, $x \in X$, Q is onto and since for all x^* in X^*, $\|Q J' x^*\| = \|x^*\|$, we have $\|Q\| = 1$. Now, let $P = J' Q$. Then $P(X^{***}) = J'(X^*)$. Finally, because $Q J' x^* = x^*$ for every x^* in X^*, we obtain $P^2 x^{***} = J'(Q J') Q x^{***} = J' Q x^{***}$ $= P x^{***}$ and P is a projection. Since $\|P\| \leqslant \|J'\| \|Q\| = 1$ and since $\|P x^{***}\| = \|x^{***}\|$, $x^{***} \in P(X^{***})$, P is of norm one.

Based on this lemma we have

Theorem 6. *If X^{**} has a retro-basis and if for every x^{**} in X^{**} there is a unique x^{****} in X^{*****} for which $\|x^{****}\| = \|x^{**}\|$ such that $x^{**}(x^*)$ $= x^{****}(J' x^*)$ for all x^* in X, then X is reflexive.*

Proof. Let $\{x_i^{**}, J' x_i^*\}$ be a retro-basis for X^{**}. Then Theorem III.2.12 implies that $\{x_i^*, x_i^{**}\}$ is a basis for X^*. Since by the preceding lemma for every X there is a projection of unit norm of X^{***} on $J'(X^*)$, the assumptions of Theorem IV.1.4 are satisfied. Hence $\sum_{i \leqslant n} x^{***}(x_i^{**}) x_i^*$ converges in X^* for all x^{***} in X^{***}. As a consequence of the foregoing theorem, X^* is reflexive. But this is equivalent to the reflexivity of X (I.3.18), and the theorem is proved.

Theorem 7. *If* X *has an unconditional basis and* X^{**} *is separable, then* X *is reflexive.*

Proof. The separability of X^{**} implies that of X^* (I.3.11). Then, on account of Corollary 1.4, X^* is weakly sequentially complete. Hence the unit ball U^* in X^* is conditionally weakly sequentially complete. Let $\{y_n^*\}$ be a sequence in U^*, weakly converging to a point y^* of X^*. Then there is a y^{**} in X^{**}, $\|y^{**}\|=1$, such that. $\|y^*\|=y^{**}(y^*)$ (I.3.10). But $y^{**}(y^*)=\lim_n y^{**}(y_n^*)\leqslant\sup_n\|y_n^*\|\leqslant 1$ implies that U^* is weakly sequentially complete. This fact, combined with the hypothesis, yields (I.3.19) that X^* and hence X is reflexive.

Theorem 8. *If* X *has an unconditional basis and if no subspace of* X *is topologically isomorphic to* l_1, *then every bounded set in* X *is sequentially compact in the weak topology for* X.

Proof. We take an arbitrary bounded sequence $\{y_n\}$ in X and define $\{x_i\}$ to be an unconditional basis for X with associated biorthogonal sequence $\{x_i^*\}$. Since $\sup_n|x_i^*(y_n)|\leqslant\|x_i^*\|\sup_n\|y_n\|<\infty$ and since every bounded closed subset of Φ is sequentially compact in the usual topology for Φ, one can choose a subsequence $\{z_n\}$ of the sequence $\{y_n\}$ such that there exist the limits

$$\alpha_i=\lim_n x_i^*(z_n),\qquad i=1,2,\dots.$$

If $\{p_n\}$ and $\{q_n\}$ are arbitrary increasing sequences of positive integers, then

$$\sup_n\|z_{p_n}-z_{q_n}\|\leqslant 2\sup_n\|y_n\|<\infty$$

and

$$\lim_n x_i^*(z_{p_n}-z_{q_n})=0,\qquad i=1,2,\dots.$$

Suppose now that $\{z_{p_n}-z_{q_n}\}$ is not weakly convergent to 0. Then there is an x^* in X^* with $\|x^*\|=1$ and a subsequence $\{w_n\}$ of $\{z_{p_n}-z_{q_n}\}$ such that $\inf_n\|w_n\|\geqslant\inf_n|x^*(w_n)|>0$. The definition of $\{w_n\}$ implies $\lim_n x_i^*(w_n)=0$ for all i. Thus, by Theorem IV.3.7, a subsequence $\{v_n\}$ of $\{w_n\}$ is a basis for $\overline{\mathrm{sp}}\{v_n\}$, equivalent to a block basis with respect to the basis $\{x_i\}$. It is clear that the basis $\{v_n\}$ is unconditional, since the basis $\{x_i\}$ is unconditional.

For $\alpha=\{\alpha_i\}\in l_1$, $\lim_n\sum_{i\leqslant n}\alpha_i v_i$ exists, since $\sup_n\|v_n\|\leqslant\sup_n\|w_n\|$ $\leqslant\sup_n\|z_{p_n}-z_{q_n}\|<\infty$, and the limit element is in $\overline{\mathrm{sp}}\{v_n\}$. On the other hand, if $\sum_{i\leqslant n}\alpha_i v_i$ is convergent, since the basis $\{v_n\}$ is unconditional,

we have $\sum_{i=1}^{\infty} |\alpha_i| |x^*(v_i)| < \infty$ as a result of Riemann's theorem (II.2.2).
But because $\inf_n |x^*(v_n)| > 0$ this is true only if α is in l_1. Therefore we
can define the linear transformation T of l_1 onto $\overline{\mathrm{sp}}\{v_n\}$, by $T\alpha$
$= \lim_n \sum_{i \leq n} \alpha_i v_i, \alpha \in l_1$. Evidently $\|T\| \leqslant \sup_n \|v_n\| < \infty$. The fact that $\{v_n\}$
is a basis for $\overline{\mathrm{sp}}\{v_i\}$ and the assumption $T\alpha = 0$ for any α in l_1 imply
$\alpha_i = 0$ for all i, hence that α is the zero element of l_1. Thus T is one-to-one,
and, by (I.2.6), is a topological isomorphism. That a subspace of X is
topologically isomorphic with l_1 contradicts our hypothesis, thus
$\{z_{p_n} - z_{q_n}\}$ converges weakly to 0.

Therefore the sequence $\{z_n\}$ converges in the weak topology for X,
because otherwise there would exist a y^* in X^* and sequences $\{p_n\}$
and $\{q_n\}$ of increasing integers such that $\inf_n |y^*(z_{p_n} - z_{q_n})| > 0$ which,
by the above results, is impossible. Thus every bounded sequence in X
contains a subsequence which converges in the weak topology for X
and the proof of the theorem is complete.

Theorem 9. *In every separable non-reflexive Banach space X there
exists a subspace with a non-shrinking basis.*

Proof. By (I.3.18), the unit ball U of X can not be weakly sequentially
compact. We thus have two possibilities: (i) X is not weakly sequentially
complete. In this case there exists a weakly convergent sequence $\{x_n\}$
in X which has no weak limit in X. (ii) X is weakly sequentially complete
but U is not weakly sequentially compact. If this is the case, there is a
sequence $\{x_n\}$ in U such that no subsequence of $\{x_n\}$ is a weak Cauchy
sequence.

(i) Without loss of generality one may take $x_1 = 0$. Let $J(J')$ be the
natural embedding of $X(X^*)$ into X^{**} (into X^{***} respectively). Since
$\sup_n \|x_n\| < \infty$ (I.3.15), we can define $x_1^{**} \in X^{**}$ by $x_1^{**}(x^*) = \lim_n x^*(x_n)$,
$x^* \in X^*$. Obviously $x_1^{**} \notin J(X)$. Next, let $x_n^{**} = x_1^{**} - J x_n$, $n = 2, 3, \ldots$.
It follows that $\sup_n \|x_n^{**}\| \leqslant \|x_1^{**}\| + \sup_n \|x_n\| < \infty$ and, since $x_1^{**} \notin J(X)$
that $\inf_n \|x_n^{**}\| = \inf_n \|x_1^{**} - J x_n\| > 0$. Since $\|x^{**}\| = \sup\{|x^{**}(x^*)| \, | \, \|x^*\| \leqslant 1,$
$x^* \in X^*\} = \sup\{|x^{***}(x^{**})| \, | \, \|x^{***}\| \leqslant 1, x^{***} \in J'(X^*)\}$, $J'(X^*)$ is a determining manifold for X^{**}. Moreover, $\lim_n J'x^*(x_n^{**}) = x_1^{**}(x^*) - \lim_n J x_n(x^*)$
$= x_1^{**}(x^*) - \lim_n x^*(x_n) = 0$ for all x^* in X^*. Thus Theorem IV.1.6 implies
the existence of a subsequence $\{x_{n_k}^{**}\}$ of $\{x_1^{**}, x_2^{**}, x_3^{**}, \ldots\}$ with $n_1 = 1$,
which is a basis for $\overline{\mathrm{sp}}\{x_{n_k}^{**}\} = \overline{\mathrm{sp}}\{x_1^{**}, x_1^{**} - J x_{n_2}, x_1^{**} - J x_{n_3}, \ldots\}$
$= \overline{\mathrm{sp}}\{x_1^{**}, J x_{n_2}, J x_{n_3}, \ldots\}$.

Let now $Y = \overline{\mathrm{sp}}\{x_{n_2}, x_{n_3}, \ldots\}$ and $Z = \overline{\mathrm{sp}}\{x_{n_{k+1}}^{**}\} = \overline{\mathrm{sp}}\{x_1^{**} - J x_{n_2},$
$x_1^{**} - J x_{n_3}, \ldots\}$. Furthermore, let I' be the identity in $B(\overline{\mathrm{sp}}\{x_{n_k}^{**}\})$ and

let U_1 be the first expansion operator of the basis $\{x^{**}_{n_k}\}$. U_1 is a continuous projection (III.2.3) with one dimensional range spanned by x^{**}_1. We then define the bounded linear transformation $T: Y \to Z$ by $T y = (I' - U_1) J y$, $y \in Y$. T is one-to-one since for $y \in Y$, $0 = T y = (I' - U_1) J y$ implies that $J y$ is in the range of U_1 which is possible only if $y = 0$. T is onto, since any $z \in Z$ has the unique representation as a sum $z = \alpha x^{**}_1 + J y$, $\alpha \in \Phi$, $y \in Y$ (according to $x^{**}_1 \notin J(Y)$ one has $\alpha x^{**}_1 = J y = 0$ whenever $\alpha x^{**}_1 + J y = 0$) so that $z - \alpha x^{**}_1 \in J(Y)$ and $T J^{-1}(z - \alpha x^{**}_1) = (I' - U_1)(z - \alpha x^{**}_1) = z$. It thus follows from (I.2.6) that T is a topological isomorphism of Y onto Z.

Moreover, by (IV.1.5) it is clear that $\{T^{-1} x^{**}_{n_{k+1}}\}$ is a basis for Y, which is non-shrinking as we will show by the following argument: If $\{T^{-1} x^{**}_{n_{k+1}}\}$ would be shrinking, then this would imply that $\lim_k y^*(T^{-1} x^{**}_{n_k}) = 0$ for all $y^* \in Y^*$, hence that $\lim_k z^*(x^{**}_{n_k}) = \lim_k T^* z^*(T^{-1} x^{**}_{n_k}) = 0$ for every $z^* \in Z^*$. Since the restriction to Z of every x^{***} in X^{***} is automatically in Z^* and since then $\lim_k x^{***}(J x_{n_k}) = x^{***}(x^{**}_1)$, $x^{***} \in X^{***}$, we have by (I.3.15) the result that $x^{**}_1 \in J(X)$ which is impossible. This shows that under assumption (i), $\{T^{-1} x^{**}_{n_{k+1}}\}$ is a non-shrinking basis for the subspace Y of X.

(ii) Let $\{g_m\}$ be a countable dense set of non-zero elements in X. For every m there is a $g^*_m \in X^*$ with $\|g^*_m\| = 1$ and $g^*_m(g_m) = \|g_m\|$ (I.3.10). Then $\Gamma = \overline{\mathrm{sp}}\{g^*_m\}$ in X^* is a determining manifold for X, since for $x \in X$, $\sup\{|x^*(x)| \mid \|x^*\| \leqslant 1, x^* \in \Gamma\} \leqslant \|x\| \leqslant \inf_m(\|x - g_m\| + |g^*_m(x)| + |g^*_m(g_m - x)|)$ $\leqslant \sup_m |g^*_m(x)| \leqslant \sup\{|x^*(x)| \mid \|x^*\| \leqslant 1, x^* \in \Gamma\}$. Next, the following diagonal procedure gives a subsequence $\{y_n\}$ of $\{x_n\}$ such that $\lim_n g^*_m(y_n)$ exists for $m = 1, 2, \ldots$. Since $x^*_1(U)$ is sequentially compact in Φ (I.1.12 and 9), there is a subsequence $\{x^{(1)}_n\}$ of $\{x_n\}$ such that $\lim_n g^*_1(x^{(1)}_n)$ exists and we take $y_1 = x^{(1)}_1$. Similarly, there is a subsequence $\{x^{(2)}_n\}$ of $\{x^{(1)}_2, x^{(1)}_3, \ldots\}$ such that $\lim_n g^*_2(x^{(2)}_n)$ exists and we take $y_2 = x^{(2)}_1$. Continuing in this way it is easy to see that $\{y_n\}$ has the requested properties. Since $\{y_n\}$ can not be a weak Cauchy sequence, there is an $x^* \in X^*$ of norm one and there are increasing sequences of integers $\{p_k\}$ and $\{q_k\}$ such that $\inf_k |x^*(y_{p_k} - y_{q_k})| > 0$. The sequence $\{z_n\} = \{y_{p_n} - y_{q_n}\}$ fulfils the hypothesis of Theorem IV.1.6, because $0 < \inf_n |x^*(z_n)| \leqslant \inf_n \|z_n\|$ $\leqslant \sup_n \|z_n\| \leqslant 2 \sup_n \|y_n\| \leqslant 2$ and $\lim_k g^*_m(z_n) = \lim_k g^*_m(y_{p_k} - y_{q_k}) = 0$, $m = 1, 2, \ldots$ (we observe that, given $g^* \in \Gamma$ and $\varepsilon > 0$, by the last inequalities and by $\Gamma = \overline{\mathrm{sp}}\{g^*_m\}$, there is an m such that $\|g^* - f^*_m\| < \varepsilon$, where f^*_m is in $\mathrm{sp}\{g^*_1, \ldots, g^*_m\}$, and an n_0 depending on m such that $|f^*_m(z_n)| < \varepsilon$, $n \geqslant n_0$. Hence $|g^*(z_n)| \leqslant 2 \|g^* - f^*_m\| + |f^*_m(z_n)| < 3\varepsilon$, $n \geqslant n_0$ and one has

$\lim\limits_{n} g^*(z_n)=0$ for all $g^* \in \Gamma$). Thus a subsequence $\{z_{n_k}\}$ of $\{z_n\}$ forms a basis for $\overline{sp}\{z_{n_k}\}$. Finally, due to $\inf\limits_{k}|x^*(z_{n_k})|>0$, this basis is not shrinking and the theorem now follows.

It is known that a Banach space X is reflexive if and only if every separable closed linear subspace of X is reflexive. We can now prove the following stronger proposition:

Theorem 10. (PELCZYNSKI) *A Banach space X is reflexive if and only if every subspace of X with a basis is reflexive.*

Proof. By (I.3.18) and Theorem 1.1, the necessity is obvious. On the other hand, let us assume that X is not reflexive. Then, again by (I.3.18), there would be a separable closed linear subspace Y of X which is not reflexive, since otherwise X itself would be reflexive. Furthermore, the preceding theorem would imply the existence of a subspace Z of Y with a non-shrinking basis. But by hypothesis Z must be reflexive so that by Theorem 2 the basis for Z must be shrinking. This contradiction shows that X is reflexive and we are done.

The next corollary is immediate:

Corollary 11. *In every non-reflexive Banach space there exists a subspace with a non-shrinking basis.*

3. Criteria for Finite Dimension

We observe that if X is infinite dimensional and has an absolutely convergent basis, then, as a result of Theorem III.5.2, X is topologically isomorphic with l_1. Thus, as a corollary of this theorem we obtain

Corollary 1. *If X has an absolutely convergent basis and X^* is weakly sequentially complete, then X is finite dimensional.*

Proof. Supposing X to be infinite dimensional, then it is topologically isomorphic with l_1. Hence by (I.3.25) X^* is topologically isomorphic with l_1^* and thus with l_∞ (I.4.c). Therefore, l_∞ would also be weakly sequentially complete. This is easy to see by the following argument. Let T be a topological isomorphism of a weakly sequentially complete Banach space Y onto a Banach space Z and let $\{z_k\}$ be an arbitrary weakly converging (not necessarily in the weak topology of Z) sequence in Z. Then $\lim\limits_{k} T^* z^*(T^{-1} z_k)=\lim\limits_{k} z^*(z_k)$ exists for all z^* in Z^*. Since T^* is a topological isomorphism of Z^* onto Y^* (I.3.25), $T^{-1}(z_k)$ converges weakly in Y, to an element y of Y because Y is weakly sequentially complete. The weak sequential completeness of Z then follows from

$\lim\limits_{k} z^*(Ty-z_k)=\lim\limits_{k} T^* z^*(y-T^{-1}z_k)=0$, $z^*\in Z^*$. Finally, the result that l_∞ is weakly sequentially complete leads to a contradiction (I.4.c) so that X is of finite dimension.

Corollary 2. *Let X have an absolutely convergent basis and let X^* be separable. Then X can not be infinite dimensional.*

Proof. In just the same way as in the proof of Corollary 1 it can be shown that l_∞ would be separable under the hypothesis stated in this corollary. But this is not the case (I.4.c). Therefore, X is finite dimensional.

References for Chapter V: BESSAGA and PELCZYNSKI [3], DAY [2], GELBAUM [1], JAMES [4], KARLIN [2], PELCZYNSKI [4] and TAYLOR [1].

CHAPTER VI

Bases for Hilbert Spaces

The separable Hilbert spaces are the only class of abstract spaces for which the existence of a Schauder basis is warranted. It is indeed a classical fact that in every separable Hilbert space H there exists a total orthonormal sequence, hence an orthonormal basis for H. It follows that a basis for H is orthonormal if and only if it is normal and that such a basis is unconditional. On the other hand a basis $\{x_i, x_i^*\}$ for H is unconditional if and only if both $\sum_{i=1}^{\infty} |(x, x_i)|^2$ and $\sum_{i=1}^{\infty} |(x, x_i^*)|^2$ exist for every x in H.

1. Monotone and Orthonormal Bases

Let H be a separable Hilbert space.

Theorem 1. *There exists a monotone basis for every separable Hilbert space.*

Proof. Since H is assumed to be separable there exists a sequence $\{x_i\}$ which is a dense set in H. Defining $N_n = \mathrm{sp}\{x_i | i \leqslant n\}, n = 1, 2, \ldots$, it is clear that the sequence $\{N_n\}$ of finite dimensional subspaces of H is directed by inclusion and that $\overline{\mathrm{sp}}\{N_n\} = H$. Without loss of generality one may suppose that every finite collection of elements of $\{x_i\}$ is linearly independent. Thus, $\dim N_n = n$. Moreover, the orthogonal projections of H on N_n all have norm one. Consequently, H is a simple $\mathcal{N}_1$-space and H has a monotone basis by Theorem IV.2.5.

Remark. As a consequence of Corollary III.2.9, $\{x_i, x_i^*\}$ is a basis for H if and only if $\{x_i^*, x_i\}$ is a basis for H.

Definition 2. *A basis $\{x_i, x_i^*\}$ for H is called orthonormal if $x_i^* = x_i$ for all i, normalized if $\|x_i\| = 1$ for all i, and normal if $\|x_i^*\| = \|x_i\| = 1$ for all i.*

In this chapter we use the well known fact that the transformation $T:H\to H^*$ defined by $Ty(x)=(x,y)$, $x,y\in H$, is a one-to-one additive isometric map of H onto H^* and that $T(\alpha y)=\bar{\alpha}Ty$, $\alpha\in\Phi$, $y\in H$. We customarily denote the elements Ty and y by the same symbol. It is evident that an orthonormal basis is *normal*, since then $\|x_i\|^2=(x_i,x_i)$ $=(x_i,x_i^*)=x_i^*(x_i)=1$ and in the same way one obtains $\|x_i^*\|=1$.

Theorem 3. *An orthonormal basis for H is monotone.*

Proof. Let $U_n x=\sum_{i\leqslant n}(x,x_i)x_i$. Then $\|U_n x\|^2=\sum_{i\leqslant n}|(x,x_i)|^2$ and this is a non-decreasing function of n for every x in X.

We now have the following classical theorem:

Theorem 4. *There exists an orthonormal basis* $\{x_i\}$ *for H and Parseval's identity* $\|x\|^2=\sum_{i=1}^{\infty}|(x,x_i)|^2$ *holds for every x in H.*

Proof. H, as a separable space, has a total sequence $\{z_i\}$ of elements in H (I.1.11). Without loss of generality we may assume each finite subset of $\{z_i\}$ to be linearly independent. To this sequence we now apply an orthogonalization process, which, for sake of completeness, will be shortly described here:

Let us take $x_1=z_1/\|z_1\|$ and let us define recurrently $y_{n+1}=z_{n+1}$ $-\sum_{i=1}^{n}(z_{n+1},x_i)x_i$, $x_{n+1}=y_{n+1}/\|y_{n+1}\|$, $n=1,2,\dots$. Obviously $\|x_1\|=1$ and $(y_2,x_1)=0$. Because z_2 and x_1 are linearly independent we observe that $y_2\neq 0$. Hence $\|x_2\|=1$ and $(x_2,x_1)=0$. Repeating this argument one obtains $\|x_n\|=1$ and $(x_n,x_1)=0$, $n=2,3,\dots$. By the same procedure we can show that $(x_n,x_m)=0$, $n>m$. Therefore, $(x_n,x_m)=\delta_{nm}$.

Clearly, the elements $x_1,\dots,x_n$ are linearly independent. Hence the subspaces spanned by $x_1,\dots,x_n$ and by $z_1,\dots,z_n$ are the same, showing that $\{x_i\}$ is likewise a total sequence in H. Now, if $\{\alpha_i\}$ is an arbitrary sequence in Φ, $\left\|\sum_{i\leqslant n}\alpha_i x_i\right\|^2=\sum_{i\leqslant n}|\alpha_i|^2\leqslant\sum_{i\leqslant m}|\alpha_i|^2=\left\|\sum_{i\leqslant m}\alpha_i x_i\right\|^2$ for each n,m, $n\leqslant m$. It then remains to invoke the theorem of NIKOL'SKIĬ (IV.1.5) to infer that $\{x_i\}$ is an (orthonormal) basis for H.

Finally, the continuity of the inner product shows Parseval's identity, i.e. it shows that $\|x\|^2=\left(\lim_n\sum_{i\leqslant n}(x,x_i)x_i,x\right)=\lim_n\sum_{i\leqslant n}(x,x_i)(x_i,x)$ $=\sum_{i=1}^{\infty}|(x,x_i)|^2$.

Corollary 5. *Every separable infinite dimensional Hilbert space H is isometrically isomorphic to* l_2.

Proof. By Theorem 4 there exists an orthonormal basis $\{x_i\}$ for H. Since for every x in H, by the same theorem, $\|x\|^2 = \sum_{i=1}^{\infty} |(x, x_i)|^2$, the linear transformation $T: H \to l_2$ defined by $Tx = \{(x, x_i)\}, x \in H$, is an isometry (i.e. $\|Tx\| = \|x\|, x \in H$) into l_2. Hence T is one-to-one and we prove that T is onto: Let $\{\alpha_i\}$ be an arbitrary element in l_2. Then the series $\sum_{i=1}^{\infty} \alpha_i x_i$ converges to some x in H, because H is complete and $\left\| \sum_{i=p}^{q} \alpha_i x_i \right\|^2$
$= \sum_{i=p}^{q} |\alpha_i|^2$. But since $\{x_i\}$ is a basis for H, by uniqueness, we have $\alpha_i = (x, x_i)$ for all i. Therefore, $\{\alpha_i\}$ is in $T(H)$ and the theorem now follows.

Definition 6. *The unit ball U of a Banach space X is strictly convex provided that $\|tx + (1-t)y\| < 1$ whenever $x, y \in U$, $\|x\| = \|y\| = 1$, $x \neq y$ and $0 < t < 1$.*

Lemma 7. *If the unit ball of the conjugate X^* of a Banach space X is strictly convex and if Z is an arbitrary proper closed linear subspace of X, then for each z^* in Z^* there is a unique norm preserving extension x^* in X^*.*

Proof. We suppose to have two norm preserving extensions x^* and y^* to the whole space X (at least one such extension exists by the Hahn-Banach theorem (I.3.9)). Then $z^*(z) = (tz^* + (1-t)z^*)(z) = (tx^* + (1-t)y^*)(z)$, $z \in Z$ and hence $\|z^*\| \leqslant \|tx^* + (1-t)y^*\|$. But if $0 < t < 1$ and $x^* \neq y^*$, one has $\|tx^* + (1-t) y^*\|/\|z^*\| < 1$ which contradicts the foregoing inequality. Hence we must conclude that $x^* = y^*$ as we wished to prove.

Corollary 8. *If Z is an arbitrary proper closed linear subspace of H, then for each z^* in Z^* there is a unique norm preserving linear extension on H.*

Proof. One has only to show that the unit ball U of H is strictly convex. Let $x, y \in U$, $\|x\| = \|y\| = 1$, $x \neq y$ and $0 < t < 1$. Then $\|tx + (1-t)y\|^2 = t^2 + (1-t)^2 + 2(t-t^2)\operatorname{Re}(x, y) = 1 - 2(t-t^2) + 2(t-t^2)\operatorname{Re}(x, y) = 1 - (t-t^2)[2 - 2\operatorname{Re}(x, y)] = 1 - (t-t^2)\|x - y\|^2 < 1$.

Theorem 9. *A basis for H is orthonormal if and only if it is normal.*

Proof. Let $\{x_i, x_i^*\}$ be a basis for H such that $\|x_i\| = \|x_i^*\| = 1$ for all i. Moreover, let j be an arbitrary index and let P_j be the orthogonal projection on the one-dimensional subspace M_j of H, spanned by x_j. On account of (I.3.23), P_j^* is a topological isomorphism of M_j^* onto

$P_j^{-1}(0)^\perp$. Next, it is clear by Corollary III.2.9 that $M_j^* = \mathrm{sp}(x_j^{*'})$, where $x_j^{*'}$ is the restriction of x_j^* to M_j. It follows $\|x_j^{*'}\| = |x_j^{*'}(x_j)| = |x_j^*(x_j)| = 1 = \|x_j^*\|$. Moreover, let $y_j^* \in P_j(0)^\perp$ be defined by $y_j^* = P_j^* x_j^{*'}$. Then $\|y_j^*\| \leqslant \|P_j^*\| \|x_j^{*'}\| = \|x_j^{*'}\|$. But $\|x_j^{*'}\| = |x_j^{*'}(x_j)| = |x_j^{*'}(P_j x_j)| = |P_j^* x_j^{*'}(x_j)| = |y_j^*(x_j)| \leqslant \|y_j^*\|$ implies that $\|y_j^*\| = \|x_j^*\|$. Since P_j is orthogonal, $P_j^{-1}(0) = M_j^\perp$. Thus $P_j^{-1}(0)^\perp = M_j^{\perp\perp} = M_j$ (I.1.16) and the equations $y_j^*(x_j) = P_j^* x_j^{*'}(x_j) = x_j^{*'}(P_j x_j) = x_j^{*'}(x_j) = x_j^*(x_j)$ show that y_j^* is an extension of $x_j^{*'}$ to H. Hence both x_j^* and y_j^* are norm preserving extensions of $x_j^{*'}$ to H. By the preceding corollary norm preserving extensions to H of $x_j^{*'}$ are unique. Thus $y_j^* = x_j^*$ and because $y_j^* \in M_j$, there is an α in Φ such that $\bar{\alpha} x_j = x_j^*$. Finally, the relation $1 = x_j^*(x_j) = (x_j, \bar{\alpha} x_j) = \alpha \|x_j\|^2 = \alpha$ shows that $\alpha = 1$, hence that $x_j^* = x_j$. This implies that the basis $\{x_i, x_i^*\}$ is orthonormal and that $(x_i, x_j) = x_i^*(x_j) = \delta_{ij}$.

Since the converse of this proof is trivial (cf. below Definition 2) the theorem is proved.

Example 10. $\{\delta_i\}$ *is an orthonormal basis for the Hilbert space* l_2.

Proof. Clear by Theorem III.7.3.

Example 11. *The sequence* $\{x_n\}$, *given by* $x_n(z) = ((n+1)/\pi)^{\frac{1}{2}} z^n$, $z \in \mathbb{C}$, $|z| < 1$, $n = 0, 1, 2, \ldots$, *is an orthonormal basis for the Hilbert space* A^2 (cf. (I.4.g)).

Proof. Let D be the open unit disc in the complex plane $\mathbb{C}$, $D = \{z \mid |z| < 1\}$. That $\{x_n\}$ is an orthonormal sequence is proved by evaluating

$$(x_n, x_m) = \pi^{-1}(n+1)^{\frac{1}{2}}(m+1)^{\frac{1}{2}} \int_D z^n \bar{z}^m d\mu(z)$$

$$= \pi^{-1}(n+1)^{\frac{1}{2}}(m+1)^{\frac{1}{2}} \int_0^1 \int_0^{2\pi} e^{i(n-m)\varphi} \rho^{n+m} \rho \, d\rho \, d\varphi$$

$$= \pi^{-1}(n+1)^{\frac{1}{2}}(m+1)^{\frac{1}{2}}(n+m+2)^{-1} 2\pi \delta_{nm} = \delta_{nm}.$$

Now, let f be an arbitrary element in A^2. Since the Taylor series $\sum_{n=0}^{\infty} \alpha_n z^n$ for $f(z)$ converges absolutely and uniformly in each closed disc $D_r = \{z \mid |z| \leqslant r\}$, with $r < 1$, one gets

$$\int_{D_r} f(z) \overline{x_n(z)} d\mu(z) = \sum_{m=0}^{\infty} \alpha_m ((n+1)/\pi)^{\frac{1}{2}} \int_0^r \int_0^{2\pi} e^{i(m-n)\varphi} \rho^{n+m} \rho \, d\rho \, d\varphi$$

$$= \sum_{m=0}^{\infty} \alpha_m ((n+1)/\pi)^{\frac{1}{2}} (n+m+2)^{-1} 2\pi \delta_{nm} r^{n+m+2}$$

$$= \alpha_n (\pi/(n+1))^{\frac{1}{2}} r^{2n+2}.$$

Since $f(z)\overline{x_n(z)}$ is integrable over D one has

$$(f, x_n) = \lim_{r \to 1} \int_{D_r} f(z)\overline{x_n(z)}\, d\mu(z) = \alpha_n(\pi/(n+1))^{\frac{1}{2}}.$$

Because $\alpha_n z^n = (f, x_n)x_n(z)$, $z \in D$, the series $\sum_{n=0}^{\infty}(f, x_n)x_n(z)$ converges absolutely and uniformly to $f(z)$ in each compact subset of D. The application of Bessel's inequality shows that $\sum_{n=0}^{\infty}|(f, x_n)|^2 \leqslant \|f\|^2$. Since

$$\left\| \sum_{n=p}^{q}(f, x_n)x_n \right\|^2 = \sum_{n=p}^{q}|(f, x_n)|^2, \text{ the series } \sum_{n=0}^{\infty}(f, x_n)x_n \text{ converges strongly}$$

to an element, say g, in A^2. Moreover, since a subsequence of $\left\{ \sum_{n=0}^{m}(f, x_n)x_n(z) \right\}$ converges to $g(z)$ almost everywhere on D, one has $g = f$, and by Theorem III.2.4, $\{x_n, x_n\}$ is an (orthonormal) basis for A^2.

2. Unconditional Bases for Hilbert Spaces

Theorem 1. *An orthonormal basis for H is unconditional.*

Proof. Let $\{x_i\}$ be such a basis for H and let $\{n_i\}$ be an increasing sequence of integers. Then in view of Parseval's identity (Theorem 1.4) or in view of Bessel's inequality, for every x in H the series $\sum_{i=1}^{\infty}|(x, x_i)|^2$ is convergent. Therefore, since $\left\| \sum_{i=m}^{n}(x, x_{n_i})x_{n_i} \right\|^2 = \sum_{i=m}^{n}|(x, x_{n_i})|^2$ for each $m \leqslant n$ and since H is complete, the series $\sum_{i=1}^{\infty}(x, x_i)x_i$ is subseries, and hence unconditionally convergent (Theorem II.1.3). This finishes the proof of the theorem.

Next, let Σ be the set of all finite subsets of the set of all positive integers. We then have

Lemma 2. (ORLICZ) *Let S be a compact interval in $\mathbb{R}$ and let $\{f_i\}$ be a sequence in $L_2(S)$. If $\sup\left\{ \left\| \sum_{i \in \mu} f_i \right\| \mu \in \Sigma \right\} < \infty$, then one has $\sup_{n} \sum_{i \leqslant n} \|f_i\|^2 < \infty$.*

Proof. We make use again of Rademacher's system $\{\Psi_n\}$, defined below Theorem III.7.7. Bessel's inequality applied to $\sum_{i \leqslant n}\alpha_i\Psi_i, \alpha_i \in \Phi$, $i = 1, 2, \ldots$, yields

$$\sum_{i \leqslant n}|\alpha_i|^2 \leqslant \int_0^1 \left| \sum_{i \leqslant n}\alpha_i\Psi_i(t) \right|^2 dt.$$

This inequality implies that almost everywhere on S

$$\sum_{i \leqslant n} |f_i(s)|^2 \leqslant \int_0^1 \left| \sum_{i \leqslant n} f_i(s) \Psi_i(t) \right|^2 dt .$$

We now define the set

$$\mu_t^\pm = \{i \mid \Psi_i(t) \lessgtr 0, i=1, ..., n\}, \quad t \in [0,1] .$$

Integrating over S on both sides of the last inequality and applying the Fubini-Tonelli theorem then gives

$$\sum_{i \leqslant n} \|f_i\|^2 \leqslant \int_S \left[\int_0^1 \left| \sum_{i \leqslant n} \Psi_i(t) f_i(s) \right|^2 dt \right] ds$$

$$= \int_0^1 \left\| \sum_{i \leqslant n} \Psi_i(t) f_i \right\|^2 dt$$

$$\leqslant \int_0^1 \left[\left\| \sum_{i \in \mu_t^+} f_i \right\| + \left\| \sum_{i \in \mu_t^-} f_i \right\| \right]^2 dt$$

$$\leqslant 4 \sup \left\{ \left\| \sum_{i \in \mu} f_i \right\|^2 \Big| \mu \in \Sigma \right\} < \infty ,$$

as we wished to prove.

The following theorem is based on this lemma:

Theorem 3. *Let $\{x_i, x_i^*\}$ be a normalized basis for H. Then the basis is unconditional if and only if $\sum_{i \leqslant n} |(x, x_i)|^2$ and $\sum_{i \leqslant n} |(x, x_i^*)|^2$ converge for each x in H.*

Proof. "If part": Let x and y be arbitrary elements of H. From the hypothesis and through application of Schwarz's inequality it follows immediately that $\sum_{i=1}^{\infty} |(x, x_i^*)(y, x_i)| < \infty$. For every increasing sequence of integers $\{m_i\}$, therefore, $\lim_n \left(\sum_{i \leqslant n} (x, x_{m_i}^*) x_{m_i}, y \right)$ exists. H, as a reflexive space, is weakly sequentially complete (I.3.17). From this we infer that the series $\sum_{i=1}^{\infty} (x, x_i^*) x_i$ is subseries convergent in the weak topology of H. By Theorems II.1.2 and 3 the series is then unconditionally convergent in H. Hence $\{x_i\}$ is an unconditional basis for H.

"Only if part": Let $\{x_i, x_i^*\}$ be a normalized unconditional basis for H. Because $H^*(=H)$ is separable and reflexive, Theorem III.4.6 implies that $\{x_i^*, x_i\}$ is an unconditional basis for H. By Corollary VI.1.5, every separable Hilbert space is isometrically isomorphic to l_2 and hence also with the Hilbert space $L_2[0,1]$ (of course the spaces are all over the same field Φ). From the section below Definition III.4.1 we know that $\sup\left\{\left\|\sum_{i\in\mu}(x, x_i^*)\,x_i\right\|\,\middle|\,\mu\in\Sigma\right\}<\infty$ for every x in H. Thus, taking $f_i=(x, x_i^*)\,T\,x_i$, where T is an isometric isomorphism of H onto $L_2[0, 1]$, the assumptions of the preceding lemma are fulfilled. Hence $\sup_n \sum_{i\leqslant n} |(x, x_i^*)|^2 = \sup_n \sum_{i\leqslant n} \|(x, x_i^*)\,x_i\|^2 = \sup_n \sum_{i\leqslant n} \|f_i\|^2 < \infty$. In the same way we infer that $\sup_n \sum_{i\leqslant n} |(x, x_i)|^2 = \sup_n \sum_{i\leqslant n} |(x, x_i)\,(x_i, x_i^*)|^2$ $\leqslant \sup_n \sum_{i\leqslant n} \|(x, x_i)\,x_i^*\|^2 < \infty$, taking $f_i=(x, x_i)\,T x_i^*$, using again the foregoing lemma, and the fact that $\sup\left\{\sum_{i\in\mu}\|(x, x_i)\,x_i^*\|\,\middle|\,\mu\in\Sigma\right\}<\infty$. This completes the proof of the theorem.

References for Chapter VI: GELBAUM [1], KARLIN [2] and ORLICZ [2].

CHAPTER VII

Decompositions

It is quite natural to generalize the concept of a basis for a space X by taking a sequence of linear (not necessarily closed) subspaces of X instead of a sequence of elements of X and, in the same time, to define X to be an F-space. Such a basis of subspaces is called a decomposition and, if the subspaces are closed, a Schauder decomposition. The closure of the subspaces is intimately connected with the continuity of projections onto these subspaces. If X is a Banach space, it turns out that X always has a decomposition, but there are (non-separable) Banach spaces which do not have a Schauder decomposition. Concerning the existence of Schauder decompositions of Banach spaces, a theorem is verified which corresponds to the theorem of NIKOL'SKIĬ for the existence of a basis. Finally, it is shown that the notions of a weak Schauder decomposition and a Schauder decomposition in a Banach space are equivalent.

1. Decompositions of F-Spaces

Unless otherwise stated, X is always an F-space. The concept of a decomposition has been introduced by FAGE [1] and GRINBLYUM [5]. It has been developped and generalized to Banach and F-spaces by MC-ARTHUR and RETHERFORD [1], RETHERFORD [1], RUCKLE [2], and SANDERS [2, 3].

Definition 1. *A sequence* $\{M_i\}$ *of (not necessarily closed) linear subspaces of* X *is a (weak) decomposition of* X *if and only if for each x in* X *there exists a unique sequence* $\{x_i\}$ *such that* $x_i \in M_i$ *for all i and* $x = \lim_n \sum_{i \leqslant n} x_i$ *in X (in the weak topology for X, respectively).*

The uniqueness implies the existence of (not necessarily continuous) associated projections P_i of X on M_i such that $P_i P_j = \delta_{ij} P_j$.

Definition 2. *If all subspaces M_i of a (weak) decomposition* $\{M_i\}$ *of* X *are closed, then* $\{M_i\}$ *is called a (weak) Schauder decomposition of X.*

As will soon be seen, a Schauder decomposition of X is a decomposition of X such that each of the associated projections is continuous. For a Banach space X it is clear that every Schauder decomposition of X is a weak Schauder decomposition of X.

Lemma 3. *If* $\{M_i\}$ *is a Schauder decomposition of* X, *then for each* i *there exists a closed linear subspace* N_i *of* X *such that* $M_i \oplus N_i = X$.

Proof. Let i be fixed and set $M = M_i$, the summation symbol $\sum' = \sum\limits_{\substack{j \\ j \neq i}}$ and $N = \left\{ x \in X \,\middle|\, x = \sum\limits_{j=1}^{\infty}{}' x_j, x_j \in M_j \right\}$. Since $\{M_j\}$ is a decomposition of X, we have the unique representation $x = x_i + \sum\limits_{j=1}^{\infty}{}' x_j$ for each x in X, where the sequence $\{x_j\}$ (depending on x) is such that $x_j \in M_j$, $j = 1, 2, \ldots$. Hence $M \cup N = X$ and $M \cap N = \{0\}$, the equivalent statement to $M \oplus N = X$. It now remains to show that N is closed. For this purpose, let Y be the linear space defined by $Y = \left\{ \{x_j\} \,\middle|\, \lim\limits_{n} \sum\limits_{j \leq n} x_j \text{ exists in } X, \right.$ $\left. x_j \in M_j \text{ for all } j \right\}$. Since each sequence converging in X is bounded (I.1.3), we can define a translation-invariant metric $\| \; \|$ on Y by the function

$$\|\{x_j\}\| = \sup_n \left\| \sum_{j \leq n} x_j \right\|, \qquad \{x_j\} \in Y,$$

where $\|x\|$ is the quasi-norm of x in X. It is easy to see that Lemma I.1.4 implies that the metric $\| \; \|$ on Y is a quasi-norm on Y. By the following argument, Y is complete and hence an F-space: Let for every $\varepsilon > 0$ be an n_ε such that $\|\{x_{pj} - x_{qj}\}\| < \varepsilon$ for any $p, q \geq n_\varepsilon$. Then we obtain $\sup\limits_n \|x_{pn} - x_{qn}\| \leq 2 \sup\limits_n \left\| \sum\limits_{j \leq n} (x_{pj} - x_{qj}) \right\| = 2 \|\{x_{pj} - x_{qj}\}\| < 2\varepsilon$, and since each M_j is complete (a closed subset of a complete metric space obviously is complete) there exists a sequence $\{x_{0j}\}$ in X such that for all j, $\lim\limits_{p} x_{pj} = x_{0j} \in M_j$. Moreover $\{x_{0j}\} = \lim\limits_{p} \{x_{pj}\}$ in the topology of Y, because for each n there exists a $p_n \geq n_\varepsilon$ such that $\left\| \sum\limits_{j \leq n} (x_{0j} - x_{p_n j}) \right\| < \varepsilon$, which implies that $\|\{x_{0j} - x_{pj}\}\| = \sup\limits_n \left\| \sum\limits_{j \leq n} (x_{0j} - x_{pj}) \right\| \leq \sup\limits_n \left\| \sum\limits_{j \leq n} (x_{0j} - x_{p_n j}) \right\|$ $+ \sup\limits_n \left\| \sum\limits_{j \leq n} (x_{p_n j} - x_{pj}) \right\| < 2\varepsilon$ for all $p \geq n_\varepsilon$. Since $\{x_{n_\varepsilon j}\} \in Y$ there is an $m_\varepsilon \geq n_\varepsilon$ for which $\left\| \sum\limits_{j=n}^{m} x_{n_\varepsilon j} \right\| < \varepsilon$ for all $n, m \geq m_\varepsilon$. Thus

$$\left\| \sum_{j=n}^{m} x_{0j} \right\| \leq \left\| \sum_{j=n}^{m} (x_{0j} - x_{n_\varepsilon j}) \right\| + \left\| \sum_{j=n}^{m} x_{n_\varepsilon j} \right\|$$

$$< 2 \sup_n \left\| \sum_{j \leq n} (x_{0j} - x_{n_\varepsilon j}) \right\| + \varepsilon < 5\varepsilon$$

for all $n, m = m_\varepsilon$. Hence $\lim_n \sum_{j \leq n} x_{0j}$ exists in X, we have $\{x_{0j}\} \in Y$ and Y is complete.

We now define the linear transformation $T: Y \to X$ by $T(\{x_j\}) = \lim_n \sum_{j \leq n} x_j$. By uniqueness the last series vanishes if and only if $\{x_j\} = 0$, hence T is one-to-one. That T is onto follows immediately from its definition and the fact that $\{M_j\}$ is a decomposition of X. Next we prove that T is a topological isomorphism: $\|T(\{x_j\})\| = \left\| \sum_{j=1}^{\infty} x_j \right\| \leq \sup_n \left\| \sum_{j \leq n} x_j \right\| = \|\{x_j\}\|$. Therefore, by (I.2.7), T is continuous and this shows that T^{-1} is also continuous (I.2.6).

Finally, let y_0 be the limit of a convergent sequence $y_1, y_2, \ldots$ in N, where $y_0 = \sum_{j=1}^{\infty} y_{0j}$, $y_{0j} \in M_j$, and where $y_k = \sum_{j=1}^{\infty\prime} y_{kj}$, $y_{kj} \in M_j$, $y_{ki} = 0$, $k = 1$, $2, \ldots$. Using the fact that T is a topological isomorphism, we conclude that $\{y_{0j}\} = T^{-1} y_0 = T^{-1} \lim_k y_k = \lim_k \{y_{kj}\}$. But from the definition of the metric in Y it follows immediately that $y_{0i} = \lim_k y_{ki} = 0$, hence that $y_0 \in N$. Therefore N is closed and the proof of the lemma is complete.

Corollary 4. *If* $\{M_i\}$ *is a weak Schauder decomposition of a Banach space* X, *then for each* i *there exists a closed linear subspace* N_i *of* X *such that* $M_i \oplus N_i = X$.

Proof. The corollary can be verified with almost the same formalism used to prove the lemma. But now, Y is defined by weakly convergent series, and by Theorem I.3.15 one can use the same metric on Y. The same theorem makes clear that $\{z_p\}$ is a Cauchy sequence in X, where $z_p = \text{weak} - \lim_n \sum_{j \leq n} x_{pj} \left(\|z_p - z_q\| \leq \sup_n \left\| \sum_{j \leq n} (x_{pj} - x_{qj}) \right\| \right)$. Since X is complete it is then possible to define $z_0 = \lim_p z_p$ and one can show that $z_0 = \text{weak} - \lim_n \sum_{j \leq n} x_{0j}$: Given $x^* \in X^*$ of norm one, there is a fixed $p \geq n_\varepsilon$ such that $\|z_0 - z_p\| < \varepsilon$ and an m_ε (depending on p) such that $\left| x^* \left(z_p - \sum_{j \leq n} x_{pj} \right) \right| < \varepsilon$, $n \geq m_\varepsilon$. Hence

$$\left| x^* \left(z_0 - \sum_{j \leq n} x_{0j} \right) \right| \leq \|z_0 - z_p\| + \left| x^* \left(z_p - \sum_{j \leq n} x_{pj} \right) \right| + \sup_n \left\| \sum_{j \leq n} (x_{pj} - x_{0j}) \right\| < 4\varepsilon$$

for all $n \geq m_\varepsilon$. Defining T by a weak limit, the rest of the proof applies without modification.

Theorem 5. *In an F-space* X *the following three conditions are equivalent:*

(i) *There is a sequence $\{M_i\}$ of closed linear subspaces of X which is a Schauder decomposition of X.*

(ii) *There is a sequence $\{P_i\}$ of continuous projections of X such that $P_i P_j = \delta_{ij} P_j$ and $\lim_n \sum_{i \leq n} P_i x = x$ for each x in X.*

(iii) *There is a sequence $\{Q_n\}$ of continuous projections of X such that $Q_m Q_n = Q_{\min(m,n)}$ and $\lim_n Q_n x = x$ for each x in X.*

The subspaces M_i and the projections P_i and Q_i are related by $P_i(X) = M_i$ and $Q_n = \sum_{i \leq n} P_i, n = 1, 2, \ldots$.

Proof. Let (i) be given. We define the linear transformations P_i of X onto M_i by $P_i x = x_i$, $i = 1, 2, \ldots$, where $\lim_n \sum_{i \leq n} x_i$ is the unique series expansion for $x \in X$. Evidently the P_i's are (not necessarily continuous) projections and $P_i P_j = \delta_{ij} P_j$. By the following reasoning, uniqueness implies the closure of P_i for each i: By Lemma 3 there exists a closed linear subspace N_i such that $M_i \oplus N_i = X$. Let y_v be a sequence in X converging to y such that $P_i y_v$ converges to z in X. Then $\lim_v (y_v - P_i y_v) = y - z$. Since $P_i y_v \in M_i$ and $y_v - P_i y_v \in N_i$, it follows that $z \in M_i$ and $y - z \in N_i$. Thus $P_i y = P_i z = z$ and, therefore, P_i is closed. By the closed graph theorem (I.2.10), the P_i's then are continuous and (ii) is proved.

Conversely, if (ii) holds, let $M_i = P_i(X)$ for all i. Since P_i is continuous, M_i is closed. For every x in X we have $x = \lim_n \sum_{i \leq n} P_i x$. To show the uniqueness of the sequence $\{P_i x\}$ we assume a sequence $\{y_i\}$ such that $\lim_n \sum_{j \leq n} y_j = 0$ and $y_i \in M_i$ for all i. But then, by continuity of each P_i, $0 = P_i \lim_n \sum_{j \leq n} y_j = \lim_n \sum_{j \leq n} P_i y_j = P_i y_i = y_i$. Hence (i) is satisfied.

Given (ii), taking $Q_n = \sum_{i \leq n} P_i$, we have only to verify that $Q_m Q_n = \sum_{i \leq m} P_i \sum_{j \leq n} P_j = \sum_{i = \min(m,n)} P_i = Q_{\min(m,n)}$ and (iii) follows.

Finally, from (iii) with $P_1 = Q_1, P_i = Q_i - Q_{i-1}, i = 2, 3, \ldots$, we have $P_i P_j = (Q_i - Q_{i-1})(Q_j - Q_{j-1}) = Q_{\min(i,j)} - Q_{\min(i,j-1)} - Q_{\min(i-1,j)} + Q_{\min(i-1,j-1)} = \delta_{ij}(Q_j - Q_{j-1}) = \delta_{ij} P_j$, which proves (ii), and where, for convenience, we have taken $Q_0 = 0$. This finishes the proof of the theorem.

As we have done in the case of bases, we sometimes denote a decomposition $\{M_i\}$ of X by $\{M_i, P_i\}$ or, since $M_i = P_i(X)$, by $\{P_i(X), P_i\}$.

Theorem 6. *Let X be an F-space and $\{Q_n\}$ a sequence of continuous projections of X such that $Q_n Q_m = Q_{\min(n,m)}$ and $\overline{\mathrm{sp}} \bigcup_{n=1}^{\infty} Q_n(X) = X$. Then $\lim_n Q_n x = x$ for each x in X if and only if $\{Q_n x\}$ is a bounded sequence in X for each x in X.*

Proof. The necessity is obvious from (I.1.3) and the convergence of $\{Q_n x\}$.

On the other hand, if x is in the set $Y = \{y | y \in Q_m(X), m < \infty\}$, then for some $m < \infty$ we have $x \in Q_m(X)$ so $x = Q_m x$. Since $\lim_n Q_n x$ $= \lim_n Q_n Q_m x = Q_m x = x$ for all x in $Q_m(X)$, $\lim_n Q_n x = x$ for all x in Y. By hypothesis, $\{Q_n x\}$ is a bounded sequence for each x in X. Hence by (I.2.8), the limit $Qx = \lim_n Q_n x$ exists for each x in X and Q is continuous and linear. But since $Qx = x$ on Y and $\bar{Y} = X$, $Qx = x$ on X and the theorem now follows.

2. Decompositions of Banach Spaces

Throughout this section X is a Banach space.

Theorem 1. *Every infinite dimensional Banach space X has a decomposition.*

Proof. According to Corollary IV.3.10 there exists a closed linear subspace Y of X with a basis. Let P be a projection of X on Y. Such a projection exists since we do not assume that P is continuous. Let now M_1 be the nullspace $P^{-1}(0) = \{x \in X | Px = 0\}$ of P. If $\{y_i\}$ is a basis for Y we define the other subspaces M_i, $i = 2, 3, \ldots$ to be the linear subspaces spanned by y_{i-1}, $i = 2, 3, \ldots$ respectively. Thus $\{M_i\}$ is a decomposition of X. However, $\{M_i\}$ must not necessarily be a Schauder decomposition of X as we see in the subsequent Corollary 3.

Lemma 2. *There is no continuous projection of l_∞ on its subspace c_0.*

Proof. Let $\{x_n^*\} \subset c_0^*$ be a sequence such that $\|x_n^*\| = 1$, $n = 1, 2, \ldots$ and such that $\lim_n x_n^*(x) = 0$ for each x in c_0. To see that such a sequence exists we have only to take $x_n^* = \delta_n \in c_0^* (= l_1)$, where $\delta_n = \{\delta_{ni}\}$. If there would be a continuous projection P of l_∞ on c_0, then $\lim_n x_n^*(Py) = 0$ for each y in l_∞. Hence $\{P^* x_n^*\}$ would be a sequence in $c_0^{***} (= l_\infty^*)$ which converges to 0 in the weak* topology of c_0^{***}. If J is the natural (isometrically isomorphic) embedding of c_0 into c_0^{**}, then by Phillips' lemma (I.4.1) it follows that $\lim_n \|J^* P^* x_n^*\| = 0$. Since $J^* P^* x_n^* \in c_0^*$ we have

$$
\begin{aligned}
\|J^* P^* x_n^*\| &= \sup \{|J^* P^* x_n^*(x)| \mid x \in c_0, \|x\| \leqslant 1\} \\
&= \sup \{|x_n^*(PJx)| \mid x \in c_0, \|x\| \leqslant 1\} \\
&= \sup \{|x_n^*(x)| \mid x \in c_0, \|x\| \leqslant 1\} \\
&= \|x_n^*\| = 1,
\end{aligned}
$$

and this contradiction verifies the lemma.

It will turn out that the following corollary is only a special case of Dean's theorem (2.13):

Corollary 3. *There exists a decomposition of the Banach space l_∞ which is not a Schauder decomposition.*

Proof. In the proof of the preceding theorem we take $X = l_\infty$ and $P(X) = Y = c_0$. By Theorem III.7.1 there is a basis $\{\delta_i\}$ for c_0, where $\delta_i = \{\delta_{ij}\}$. But from the above lemma we know that P is not continuous. Hence $M_1 = P^{-1}(0)$ is not closed, since otherwise P would be continuous. This shows that $\{M_i\}$ is not a Schauder decomposition of l_∞.

Lemma 4. *If Y is any separable closed linear subspace of l_∞ such that $c_0 \subset Y \subset l_\infty$, there is a continuous projection of Y onto c_0.*

Proof. From the separability of Y it follows the existence of a denumerable dense set in the unit ball of Y. Retaining only those elements of this set which are linearly independent and which are not in c_0, we get a set $\{x_1, x_2, \ldots\}$ in Y. We first suppose that there is only a finite number n of elements in $\{x_1, x_2, \ldots\}$. Obviously we have $c_0 \cap \operatorname{sp}\{x_i | i \leqslant n\} = \{0\}$. Let x be an element of $Y_n = c_0 \oplus \operatorname{sp}\{x_i | i \leqslant n\}$ and $x = z_0 + \sum_{i \leqslant n} \alpha_i x_i$ $(\alpha_i \in \Phi)$ its unique decomposition into components in c_0 and $\operatorname{sp} x_i, i \leqslant n$ respectively.

If $x_i = \{x_{ij}\}$, we define the (column) vectors $v_j = \{x_{1j}, \ldots, x_{nj}\}, j = 1, 2, \ldots$ These vectors are in the unit ball U_n of l_∞^n, the Banach space of all n-tuples $\{\beta_1, \ldots, \beta_n\}$ of scalars, with the corresponding norm given by the expression $\sup\{|\beta_k| \,|\, k = 1, \ldots, n\}$. Since l_∞^n is finite dimensional, U_n is sequentially compact (I.1.9 and 12). Thus there is a nonvoid set Σ in U_n, the set of all cluster points of the sequence $\{v_j\}$. Moreover, there exists for each $v = 1, 2, \ldots$ a $2/2^v$-net $\mathscr{E}_v$ with respect to U_n. By definition, there is to each v_j a vector $w_j^{(v)} = \{w_{1j}^{(v)}, \ldots, w_{nj}^{(v)}\}$ in $\mathscr{E}_v$ such that $\|v_j - w_j^{(v)}\| < 2/2^v$ and such that only a finite number of vectors in the sequence $\{w_j^{(v)}\}$ are not infinitely repeated. We then may derive from $\{w_j^{(v)}\}$ another sequence $\{\tilde{w}_j^{(v)}\}$ in $\mathscr{E}_v$ in which each vector is infinitely repeated. This can be done by taking $\tilde{w}_j^{(1)} = 0$ and by altering a finite number of vectors $w_j^{(v)}(v > 1)$ in $\{w_j^{(v)}\}$ such that for the corresponding new vectors $\tilde{w}_j^{(v)}$ we have $\tilde{w}_j^{(v)} \in \mathscr{E}_v$, $\|\tilde{w}_j^{(v)} - \tilde{w}_j^{(v-1)}\| < 6/2^v$ and $\inf\{\|\tilde{w}_j^{(v)} - v\| \,|\, v \in \Sigma\} < 2/2^v$. Furthermore, since for $v = 1$, $\sup_j \|\tilde{w}_j^{(v)} - \tilde{w}_j^{(v-1)}\| < 6/2^v$, the set of vectors $\tilde{w}_j^{(v)} = \{\tilde{w}_{1j}^{(v)}, \ldots, \tilde{w}_{nj}^{(v)}\}$ has the property that $\lim_v \tilde{w}_j^{(v)}$ exists in l_∞^n uniformly with respect to j. At last we define sequences $\{y_i^{(v)}\}$ and $\{\tilde{y}_i^{(v)}\}$ by $y_i^{(v)} = \{w_{i1}^{(v)}, w_{i2}^{(v)}, \ldots\}$ and $\tilde{y}_i^{(v)} = \{\tilde{w}_{i1}^{(v)}, \tilde{w}_{i2}^{(v)}, \ldots\}$ respectively.

Let $z^{(v)} = z_0 + \sum_{i \leqslant n} \alpha_i(y_i^{(v)} - \tilde{y}_i^{(v)})$. Since only a finite number of entries of the vector $\sum_{i \leqslant n} \alpha_i(y_i^{(v)} - \tilde{y}_i^{(v)})$ are different from zero it is clear that $z^{(v)}$ is in c_0. Since c_0 is complete, since $y_i^{(v)}$ converges for all $i \leqslant n$ to x_i with $v \to \infty$ and since $\tilde{y}_i^{(v)}$ is a Cauchy sequence for all $i \leqslant n$, $z^{(v)}$ converges to an element z in c_0.

Then

$$\|z^{(v)}\| \leqslant \left\|z_0 + \sum_{i \leqslant n} \alpha_i y_i^{(v)}\right\| + \left\|\sum_{i \leqslant n} \alpha_i \tilde{y}_i^{(v)}\right\|$$

$$\leqslant \left\|z_0 + \sum_{i \leqslant n} \alpha_i y_i^{(v)}\right\| + \left\|z^{(v)} + \sum_{i \leqslant n} \alpha_i \tilde{y}_i^{(v)}\right\|$$

$$= 2\left\|z_0 + \sum_{i \leqslant n} \alpha_i y_i^{(v)}\right\|$$

$$\leqslant 2\|x\| + 2\left\|\sum_{i \leqslant n} \alpha_i (x_i - y_i^{(v)})\right\|,$$

where we have used the fact that each entry of $\sum_{i \leqslant n} \alpha_i \tilde{y}_i^{(v)}$ is infinitely repeated (since each column vector $\tilde{w}_j^{(v)}$ is infinitely repeated).

Next, let P_n be the transformation of Y_n into c_0, which is given by $P_n x = z$, $x \in Y_n$. P_n is linear, since for $x' = z_0' + \sum_{i \leqslant n} \alpha_i' x_i$, $P_n(x + x')$ $= \lim_v \left[z_0 + z_0' + \sum_{i \leqslant n} (\alpha_i + \alpha_i')(y_i^{(v)} - \tilde{y}_i^{(v)})\right] = P_n x + P_n x'$. That P_n is a (not necessarily continuous) projection of Y_n is clear from its definition. Moreover, P_n is continuous, since $\|P_n x\| = \left\|\lim\limits_v z^{(v)}\right\| = \lim\limits_v \|z^{(v)}\| \leqslant 2\|x\|$.

Finally, if there is no finite n such that $Y_n = Y$, since $\overline{\mathrm{sp}} \bigcup\limits_{n=1}^{\infty} Y_n = Y$, there is a unique continuous extension P of the P_n's to all of Y with norm $\leqslant 2$ (I.2.12). Again, since c_0 is closed, P is a continuous projection of Y on c_0.

Remark. It is known that P has norm 2. However, for our purposes it is sufficient to know the existence of a continuous projection of Y on c_0.

Lemma 5. *Any separable Banach space having a subspace which is topologically isomorphic to c_0 admits a continuous projection onto that subspace.*

Proof. Let X be a separable Banach space, let Y be a subspace of X and let T be a topological isomorphism of Y onto c_0. If $Ty = \{z_i^*(y)\}$, $y \in Y$, each z_i^* belongs to Y^* and there is a constant $K > 0$ such that $\|z_i^*\| \leqslant K$. By the Hahn-Banach theorem (I.3.8) there is an extension $z_i^{*'} \in X^*$ of z_i^* with $\|z_i^{*'}\| = \|z_i^*\|$. Then $T' : X \to l_\infty$, given by $T'x = \{z_i^{*'}(x)\}$, $x \in X$, evidently is an extension of T to X. T' is bounded, since $\|T'x\|$ $= \sup\limits_i |z_i^{*'}(x)| \leqslant \|x\| \sup\limits_i \|z_i^{*'}\| = \|x\| \sup\limits_i \|z_i^*\| \leqslant K\|x\|$. Thus $\overline{T'(X)}$ is a separable closed linear subspace of l_∞ which, by Lemma 4, admits a continuous projection, say P', of $\overline{T'(X)}$ on c_0. Consequently, $P = T^{-1} P' T'$

is likewise continuous, and is a projection of X on Y, since $P^2 = T^{-1} P' T' T^{-1} P' T' = T^{-1} P' T' = P$.

Theorem 6. *Every separable Banach space X which has a subspace which is topologically isomorphic to c_0 has a Schauder decomposition.*

Proof. The preceding lemma shows the existence of a continuous projection P of X on the subspace Y of X which is topologically isomorphic to c_0. Hence $X = (I-P)(X) \oplus Y$. Let T be the topological isomorphism of c_0 onto Y and let $\{\delta_i\}$ be a basis for c_0 whose existence is proved in Theorem III.7.1. Since then, $\overline{sp}\{T\delta_i\} = Y$ and, using for instance Theorem IV.1.5, since $\left\| \sum_{i \leqslant n} \alpha_i T\delta_i \right\| \leqslant \|T\| \left\| \sum_{i \leqslant n} \alpha_i \delta_i \right\| \leqslant \|T\| \left\| \sum_{i \leqslant m} \alpha_i \delta_i \right\|$
$\leqslant \|T\| \|T^{-1}\| \left\| \sum_{i \leqslant m} \alpha_i T\delta_i \right\|$ for each n, m with $n \leqslant m$ and arbitrary scalars α_i, $\{T\delta_i\}$ is a basis for Y. Finally, the set $\{M_i\}$ is the desired Schauder decomposition of X, where $M_1 = (I-P)(X)$ and M_i is the one-dimensional subspace of X spanned by $T\delta_{i-1}$, $i = 2, 3, \dots$.

Theorem 7. *Let $\{M_i\}$ be a sequence of closed subspaces in X such that $\overline{sp} \bigcup_{i=1}^{\infty} M_i = X$. Then $\{M_i\}$ is a Schauder decomposition of X if and only if there exists a constant $K \geqslant 1$ such that $\left\| \sum_{i \leqslant n} x_i \right\| \leqslant K \left\| \sum_{i \leqslant m} x_i \right\|$ for all n, m with $n \leqslant m$ and for all sequences $\{x_i\}$ with $x_i \in M_i$.*

Proof. Necessity. Let $\{M_i\}$ be a Schauder decomposition of X with corresponding sequence of projections $\{P_i\}$ of X on each of the subspaces M_i. Then $x = \lim_n \sum_{i \leqslant n} P_i x$ for every $x \in X$ and the uniform boundedness principle (I.3.14) implies the existence of a constant $K \geqslant 1$ such that $\sup_n \left\| \sum_{i \leqslant n} P_i \right\| \leqslant K$. Since the projections in the set $\{P_i\}$ are mutually orthogonal it follows that $\sum_{i \leqslant n} P_i = \sum_{j \leqslant n} P_j \sum_{i \leqslant m} P_i$ for $n \leqslant m$. Therefore $\left\| \sum_{i \leqslant n} P_i x \right\| \leqslant K \left\| \sum_{i \leqslant m} P_i x \right\|$ for every $x \in X$. Setting $x_i = P_i x$, this is the necessary condition.

Sufficiency. Let $\{E_n\}$ be a sequence of closed subspaces of X, given by $E_n = \overline{sp} \bigcup_{i \leqslant n} M_i$. Starting from the obvious equation $E_1 = M_1$ we proceed recursively to show that $E_n = M_1 \oplus \cdots \oplus M_n$ (hence to show that $E_n \cap M_{n+1} = \{0\}$). We suppose that $E_n = M_1 \oplus \cdots \oplus M_n$ for some $n \geqslant 1$ and take an arbitrary $x \in E_n \cap M_{n+1}$. Clearly x then has the unique decomposition $x = \sum_{i \leqslant n} x_i$, $x_i \in M_i$. The inequality of the hypothesis implies that $\|x\| = \left\| \sum_{i \leqslant n} x_i \right\| \leqslant K \left\| \sum_{i \leqslant n} x_i - x \right\| = 0$. Thus $E_n \cap M_{n+1} = \{0\}$, implying $E_{n+1} = M_1 \oplus \cdots \oplus M_{n+1}$. Let then Q_{nm} for $m \geqslant n$ be the projection of E_m onto E_n. From the unique decomposition $x = \sum_{i \leqslant m} x_i, x_i \in M_i$, for any $x \in E_m$ and from $\|Q_{nm}x\| = \left\| \sum_{i \leqslant n} x_i \right\| \leqslant K \left\| \sum_{i \leqslant m} x_i \right\| = K\|x\|$ then follows that the norm of each of the projections Q_{nm} is bounded by K. Since for fixed n, Q_{nm} is the restriction of $Q_{n,m+1}$ to E_m for all $m \geqslant n$ and since $\overline{\mathrm{sp}}\{M_i\} = X$ it is, through extension by continuity, clear that for all of the projections Q_{nm} (with domain E_m) there exists a unique linear extension Q_n on the whole space X (I.2.12). Moreover the norm of Q_n is also bounded by K. Q_n is a projection with range E_n, since E_n is closed and since $Q_n^2 x = Q_{nn} Q_n x = Q_n x$ for all $x \in X$. Again because $\overline{\mathrm{sp}} \bigcup_{i=1}^{\infty} M_i = X$, we have for each $x \in X$ and every $\varepsilon > 0$ an index p such that $\inf \{\|x - y\| \mid y \in E_p\} < \varepsilon$. This implies $\|x - Q_n x\| \leqslant \|y - Q_n x\| + \|x - y\|$ $= \|Q_n(x - y)\| + \|x - y\| \leqslant (K+1)\varepsilon$ for $n \geqslant p$, where we have chosen a $y \in E_p$ for which $\|x - y\| \leqslant \varepsilon$. Consequently, $\lim_n \|x - Q_n x\| = 0$ for every $x \in X$.

Finally, since with $Q_0 = 0$ for each $i \geqslant 1$ one has $M_i = (Q_i - Q_{i-1})(X)$, and since for arbitrary m, n one has $Q_m Q_n = Q_{\min(m,n)}$, by Theorem 1.4, $\{M_i\}$ is a Schauder decomposition of X.

Corollary 8. *Let $\{P_i\}$ be a sequence of mutually orthogonal continuous projections of X (i.e. $P_i P_j = \delta_{ij} P_j$) such that $\overline{\mathrm{sp}} \bigcup_{i=1}^{\infty} P_i(X) = X$. Then $\{P_i(X)\}$ is a Schauder decomposition of X if and only if there exists a constant $K \geqslant 1$ such that $\sup_n \left\| \sum_{i \leqslant n} P_i \right\| \leqslant K$.*

Proof. To show the sufficiency, let $M_i = P_i(X)$ and let m, n be positive integers such that $m \geqslant n$. Since by hypothesis, $P_i P_j = \delta_{ij} P_j$, we have for $x \in X$, $\left\| \sum_{i \leqslant n} P_i x \right\| = \left\| \sum_{j \leqslant n} P_j \sum_{i \leqslant m} P_i x \right\| \leqslant K \left\| \sum_{i \leqslant m} P_i x \right\|$. Since x was arbitrary, the preceding theorem shows immediately that $\{M_i\}$ is a Schauder decomposition of X. On the other hand, if $\{P_i(X)\}$ is a Schauder decomposition of X, we have shown that there is a constant $K \geqslant 1$ for which $\sup_n \left\| \sum_{i \leqslant n} P_i \right\| \leqslant K$, and this is the necessary condition.

Corollary 9. *If X is a Banach space with a Schauder decomposition $\{P_i(X)\}$ of nontrivial subspaces of X, then each sequence $\{x_i\}$ of non-zero vectors x_i in $P_i(X)$ is itself a basis for $\overline{\mathrm{sp}}\{x_i\}$.*

Proof. The corollary follows at once from Theorem 7 and an application of Theorem IV.1.5.

Theorem 10. $\{M_i\}$ *is a weak Schauder decomposition of X if and only if it is a Schauder decomposition of X.*

Proof. Let $\{M_i, P_i\}$ be a weak Schauder decomposition of X. Given $y \in X$, let σ_n be the set $\sigma_n = \{i \leqslant n | P_i(y) \neq 0\}$ (depending on y), and let $Y = \overline{\mathrm{sp}}\{P_i y\}$. Since for all $x \in X$, $\sum_{i \leqslant n} P_i x$ converges with n weakly to x in X it is clear by (I.3.15) that $y \in Y$, and that $x^*(x) = \lim_n \sum_{i \leqslant n} x^*(P_i x)$, $x^* \in X^*$. Therefore, by (VII.1.4), (I.2.18) and (I.3.14) there exists a constant $M > 0$ such that $\sup_n \left\| \sum_{i \leqslant n} P_i \right\| \leqslant M$. Using $P_i P_j = \delta_{ij} P_j$, one obtains for every sequence $\{\alpha_i\}$ in Φ, and for $m \leqslant n$, $\left\| \sum_{i \in \sigma_m} \alpha_i P_i y \right\|$
$= \left\| \sum_{j \leqslant m} P_j \sum_{i \in \sigma_n} \alpha_i P_i y \right\| \leqslant M \left\| \sum_{i \in \sigma_n} \alpha_i P_i y \right\|$. Thus by Nikol'skiĭ's theorem (IV.1.5), $\{P_i y\}$ is a basis for Y. This implies that there are unique coefficients $\beta_i \in \Phi$ for which $y = \lim_n \sum_{i \in \sigma_n} \beta_i P_i y$ in the strong (and hence in the weak) topology of X. But by hypothesis, $\beta_i = 1$ for all i so that $\{M_i, P_i\}$ is a Schauder decomposition of X.

Conversely, if $\{M_i\}$ is a Schauder decomposition of X, then $\sum_{i \leqslant n} P_i x$ converges in the strong, hence also in the weak topology for X to x. The uniqueness is shown by the following reasoning: Suppose that $\{z_i\}$ is a sequence in X such that $z_i \in M_i$ and $\lim_n x^* \left(\sum_{i \leqslant n} z_i \right) = 0$ for all $x^* \in X^*$. Then each z_i must vanish since $x^*(z_i) = \lim_n P_i^* x^* \left(\sum_{j \leqslant n} z_j \right) = 0$ for all x^* in X^*. Thus $\{M_i\}$ is a weak Schauder decomposition of X and the proof of the theorem is complete.

Theorem 11. *Let X be reflexive. Then $\{M_i, P_i\}$ is a Schauder decomposition of X if and only if there is a Schauder decomposition of X^* with associated sequence of projections $\{P_i^*\}$.*

Proof. Let $\{M_i, P_i\}$ be a Schauder decomposition of X and let $Q_n = \sum_{i \leqslant n} P_i$. Then $x^*(x) = x^* \left(\lim_n Q_n x \right) = \lim_n x^*(Q_n x) = \lim_n Q_n^* x^*(x)$ for all x in X and all x^* in X^*. Since X is reflexive this implies that $\lim_n x^{**}(Q_n^* x^*) = x^{**}(x^*)$ for every x^{**} in X^{**} so that $Q_n^* x^*$ converges to x^* in the weak topology of X^*. Hence for each x^* in X^*, as a consequence of (I.3.15), $\sup_n \| Q_n^* x^* \| < \infty$, and x^* is in $\overline{\mathrm{sp}} \bigcup_{n=1}^{\infty} Q_n^*(X^*)$. The last property implies that $\overline{\mathrm{sp}} \bigcup_{n=1}^{\infty} Q_n^*(X^*) = X^*$. Furthermore, $Q_n^* Q_m^* = Q_{\min(m,n)}^*$, because $Q_n^* Q_m^* x^*(x) = x^*(Q_m Q_n x) = x^*(Q_{\min(m,n)} x) = Q_{\min(m,n)}^*$

$x^*(x)$. Now, by Theorem 1.6, Theorem 1.5 (iii) and by what has preceded, $\{P_i^*(X^*)\}$ is a Schauder decomposition of X^*, where $\{P_i^*\} = \{Q_1^*, Q_2^* - Q_1^*, Q_3^* - Q_2^*, \ldots\}$.

On the other hand, applying again this procedure, it follows that $\{P_i^{**}(X^{**})\}$ is likewise a Schauder decomposition of X^{**}. Let J be the natural embedding of X onto X^{**} and let $P_i = J^{-1} P_i^{**} J$. Then P_i is a continuous projection of X and $\lim_n \sum_{i \leq n} P_i x = J^{-1} \lim_n \sum_{i \leq n} P_i^{**} J x = J^{-1} J x$

$= x$ and, again by Theorem 1.5 (ii), it is clear that $\{P_i(X)\}$ is a Schauder decomposition of X.

Corollary 12. *If $\{P_i(X)\}$ is a Schauder decomposition of X and X^* has the property that the convergence of a sequence in the weak* topology of X^* implies the convergence of that sequence in the weak topology of X^*, then $\{P_i^*(X^*)\}$ is a Schauder decomposition of X^*.*

We observe that there exist Banach spaces which are not reflexive but which although have the convergence property mentioned in the above corollary. An example of such a space is the conjugate space l_∞^* of l_∞ (I.4.c). Based on this property and on the fact that each bounded linear transformation of l_∞ into a reflexive Banach space maps weak Cauchy sequences into strongly convergent sequences (I.4.c) we can now prove the following theorem, and the subsequent proposition which is used to verify the theorem.

Theorem 13. (DEAN) *The Banach space l_∞ does not have a Schauder decomposition.*

Proof. We suppose that l_∞ has a Schauder decomposition, $\{P_i(l_\infty)\}$, and show that this assumption leads to a contradiction. Without loss of generality one may suppose that each subspace $P_i(l_\infty)$ is nontrivial. We first choose a sequence $\{x_i\}$ with $x_i \in P_i(l_\infty)$ of norm one and a sequence $\{f_i\}$ with $f_i \in P_i(l_\infty)^*$, also of norm one, and such that $f_i(x_i) = 1$ (which is possible by (I.3.10)). According to the Hahn-Banach theorem each f_i has a norm-preserving extension x_i^* in l_∞^*. It follows that $P_i^* x_i^*(x_j) = x_i^*(P_i x_j) = \delta_{ij} x_i^*(x_i) = \delta_{ij} f_i(x_i) = \delta_{ij}$. Let $Y = \overline{sp}\{P_i^* x_i^*\}$ in l_∞^*. Since $\|P_i^* x_i^*\| = \sup\{|P_i^* x_i^*(x)| \,\big|\, \|x\| \leq 1, x \in l_\infty\} \geq P_i^* x_i^*(x_i) = 1$, and since by Corollary 12 $\{P_i^*(l_\infty^*)\}$ is a Schauder decomposition of l_∞^*, Corollary 9 implies that $\{P_i^* x_i^*\}$ is a basis for Y. Moreover, we then have $x^* = \lim_n \sum_{i \leq n} P_i^* x^*$ in l_∞^* for every $x^* \in l_\infty^*$. Since $|x^*(x_i)| \leq \left| \sum_{j < i} P_j^* x^*(x_i) \right| + \left\| x^* - \sum_{j < i} P_j^* x^* \right\|$ and $\sum_{j < i} P_j^* x^*(x_i) = \sum_{j < i} x^*(P_j x_i) = 0$, x_i converges with i weakly to 0 in l_∞.

Next we consider the linear transformation $T : l_\infty \to Y^*$, given by $Tx(y^*) = y^*(x)$, $x \in l_\infty$, $y^* \in Y$. T is continuous, since $\|Tx\|$

$= \sup\{|y^*(x)| \mid y^* \in Y, \|y^*\| \leqslant 1\} \leqslant \|x\|$. As we soon will show (Proposition 14), Y is reflexive. Since then Y^* is also reflexive (I.3.18) and thus weakly sequentially complete (I.3.17), this implies that T sends the weakly convergent sequence $\{x_i\}$ in l_∞, into the sequence $\{Tx_i\}$ which converges strongly in Y^* (I.4.c), say to F. But from the definition of T it follows immediately that $\lim_i Tx_i(y^*) = \lim_i y^*(x_i) = 0$ for all $y^* \in Y \subset l_\infty$, hence that Tx_i converges to 0 in the weak* topology of Y^*. Therefore, $F = 0$ so that $\lim_i \|Tx_i\| = 0$ which contradicts the estimate

$$\|Tx_i\| = \sup\{|y^*(x_i)| \mid y^* \in Y, \|y^*\| \leqslant 1\} \geqslant |P_i^* x_i^*(x_i)|/\|P_i^* x_i^*\| \geqslant 1/\sup_n \|P_n\| > 0$$

(by Corollary 8, $\sup_n \|P_n\| < \infty$). It now remains to show the following

Proposition 14. Y is reflexive.

Proof. Let $\{z_n^*\}$ be a sequence in the unit ball of Y. Since $\{P_i^* x_i^*\}$ is a basis for Y one has the unique expansion $z_n^* = \lim_m \sum_{i \leqslant m} \alpha_{ni} P_i^* x_i^*$ with $\alpha_{ni} \in \Phi$, and by (III.2.3) there is a constant $M > 0$ for which $\sup_n |\alpha_{ni}| \leqslant 2M/\|P_i^* x_i^*\|$. Thus there is an increasing sequence of positive integers $\{n_{1j}\}$ such that $\lim_j \alpha_{n_{1j}, 1}$ exists and is, say, α_1 (I.1.12 and I.1.9). Likewise, there exists a subsequence $\{n_{2j}\}$ of $\{n_{1j}\}$ such that $\lim_j \alpha_{n_{2j}, 2} = \alpha_2$. Continuing in this way (this method is called a diagonal process) one obtains a subsequence $\{z_{njj}^*\}$ such that $z_{njj}^* = \lim_m \sum_{i \leqslant m} \alpha_{njj, i} P_i^* x_i^*$ and $\lim_j \alpha_{njj, i} = \alpha_i$.

Given $\varepsilon > 0$ there is for any fixed $x \in l_\infty$ a k such that $\left\| \sum_{j=k}^\infty P_j x \right\| < \varepsilon$,

so that $|(z_{n_{pp}}^* - z_{n_{qq}}^*)(x)| \leqslant \left| \sum_{i=1}^\infty (\alpha_{n_{pp}, i} - \alpha_{n_{qq}, i}) P_i^* x_i^* \left(\sum_{j<k} P_j x \right) \right| + 2 \left\| \sum_{j=k}^\infty P_j x \right\|$

$< \left| \sum_{j<k} (\alpha_{n_{pp}, i} - \alpha_{n_{qq}, i}) P_j^* x_j^*(x) \right| + 2\varepsilon$ and this is smaller than, say 3ε, if p and q are chosen large enough. Thus $\{z_{n_{jj}}^*(x)\}$ is a Cauchy sequence in Φ for each x in l_∞ and the Banach-Steinhaus theorem implies the existence of a z^* in l_∞^* which is the weak* limit of $\{z_{n_{jj}}^*\}$ in l_∞^*. But then one has $\lim_j z_{n_{jj}}^* = z^*$ in the weak topology of l_∞ (I.4.c) and by (I.3.15) z^* is in $\overline{\mathrm{sp}}\{z_{n_{jj}}^*\} \subset Y$. Now, since every $F \in Y^*$ has an extension x_F^{**} in l_∞^{**} (I.3.9), $\lim_j F(z_{n_{jj}}^*) = \lim_j x_F^{**}(z_{n_{jj}}^*) = x_F^{**}(z^*) = F(z^*)$, $\{z_{n_{jj}}^*\}$ converges to z^* in the weak topology of Y. Finally, this shows that Y is weakly sequentially compact and hence reflexive (I.3.18).

References for Chapter VII: DEAN [1], FAGE [1], GRINBLYUM [5], MCARTHUR and RETHERFORD [1], RETHERFORD [1], RUCKLE [2], SANDERS [2, 3] and SOBCZYK [1].

CHAPTER VIII

Applications to the Theory
of Banach Algebras

In the first paragraph of this chapter we assume to have to do with Banach spaces having a basis, as is the case for separable Hilbert spaces or for most of all known concrete separable Banach spaces. Whenever this hypothesis is fulfilled we can draw conclusions on the approximation problem of compact linear operators by linear operators of finite rank.

If X is a Banach space with a basis, or at least with a Schauder decomposition, then, as a beautiful application to the theory of Banach algebras, there are many results on the properties of the corresponding proper π-rings. A π-ring is a commutative Banach algebra, which is a subset of the vector space s over the field Φ, of all sequences $\alpha = \{\alpha_i\}$ in Φ, which contains the elements $\{1, 1, 1, \ldots\}$, $\{1, 0, 0, \ldots\}$, $\{0, 1, 0, 0, \ldots\}$, $\ldots$, and whose elements satisfy some simple inequality. It turns out that the set $r = \{\alpha | \alpha \in s$ such that $\sum_{i \leqslant n} \alpha_i P_i x$ is (weakly) convergent for every x in $X\}$ is a (weak) π-ring, the proper (weak) π-ring of the decomposition $\{P_i(X), P_i\}$. Moreover, the proper weak π-rings and the proper π-rings are equivalent, semi-simple, and isometrically isomorphic with the set of operators Π in $B(X)$, defined by $A(\alpha)x = \lim_n \sum_{i \leqslant n} \alpha_i P_i x$, $x \in X$, $\alpha \in r$. Based on an existence theorem for Schauder decompositions, which is a generalization of the corresponding existence theorem for bases established by NIKOL'SKII, we derive an alternative existence theorem for Schauder decompositions of X which is related to the existence of proper π-rings.

If the proper π-rings of Schauder decompositions of Banach spaces are partially ordered in a natural way by inclusion, there exists a minimal π-ring with a corresponding minimal Schauder decomposition. The basis for a special space of sequences, w_0, ist shown to be minimal. Naturally, the properties of a proper π-ring of a basis depend on the type of the basis. Some results are established in the special case of unconditional bases.

1. Two-Sided Ideals of Operators of Finite Rank

Definition 1. *A linear transformation on a Banach space X to another Banach space Y is said to be of finite rank if its range is finite dimensional.*

Theorem 2. *If X and Y are Banach spaces and either Y has a basis or X^* has a retro-basis, then each compact linear transformation $T: X \to Y$ is approximable in norm by linear operators of finite rank.*

Proof. Let U_X be the unit ball in X and let first have Y a basis with the corresponding sequence of expansion operators $\{U_n\}$. Then $T(U_X)$ is conditionally compact and by (I.4.4) $\lim_n U_n T x = T x$, uniformly on U_X. Clearly $T_n = U_n T$ is of finite rank and the preceding consideration yields $\lim_n \|T_n - T\| = 0$.

On the other hand, let X^* have a retro-basis with the associated set of expansion operators $\{V_n^*\}$. The adjoint T^* is also compact (I.3.27), therefore $T^*(U_{Y^*})$ is conditionally compact and $\lim_n V_n^* T^* y^* = T^* y^*$ uniformly on U_{Y^*}. Hence $\lim_n \|T V_n - T\| = \lim_n \|V_n^* T^* - T^*\| = 0$. Again, $T_n = T V_n$ is of finite rank, where we note that we have used the fact that by Theorem III.2.12 each V_n^* is the adjoint of a linear operator V_n of finite rank. This completes the proof of the theorem.

Definition 3. *Let X be a Banach space, $F(X) \subset B(X)$ the set of all linear operators of finite rank of X and $\mathscr{C}(X) \subset B(X)$ the set of all compact endomorphisms of X.*

The classes $F(X)$ and $\mathscr{C}(X)$ are known to be two-sided ideals in $B(X)$, by (I.2.17) $\mathscr{C}(X)$ is closed in the uniform operator topology of $B(X)$ and for $F(X)$ we have the following corollary to Theorem 2:

Corollary 4. *Let X be a Banach space with a basis. Then $\mathscr{C}(X)$ is the closure of $F(X)$ in the uniform operator topology of $B(X)$.*

By Theorem 2 and Theorem III.2.7 the corollary remains true, if, instead of X having a basis, we only assume that X^* has a weak* Schauder basis.

2. π-Rings

We consider the complex vector space s of all sequences $\alpha = \{\alpha_i\}$ of complex numbers. Defining the product of $\alpha, \beta \in s$ by $\alpha \beta = \{\alpha_i \beta_i\}$, s becomes a commutative algebra. Moreover, it is clear that $e = \{1, 1, \ldots\}$ is the identity of s. Let $\delta_n = \{\delta_{ni}\}$, where δ_{ni} denotes Kronecker's symbol.

Lemma 1. *Each subalgebra t of s containing e and* δ_n, $n=1,2,...$, *is semi-simple.*

Proof. We define the set $u_n = \{\alpha - \delta_n \alpha | \alpha \in t\}$. It is immediately verified that u_n is an ideal of t. We show that u_n is a maximal ideal: Suppose u_n to be a proper subset of a maximal ideal u. Then we can take an element $\beta \in u$ which is not in u_n, implying $\beta_n \neq 0$. But then we have $e \in u$, since $\beta_n^{-1} \beta \in u$, $(e - \delta_n)(e - \beta_n^{-1} \beta) \in u$ and $\beta_n^{-1} \beta + (e - \delta_n)(e - \beta_n^{-1} \beta) = e$. Thus, if $e \in u$, one has $u = t$, hence u is not maximal. This contradiction shows that u_n is maximal. Since $\bigcap_n u_n = 0$, t is semi-simple and we are done.

Now, if as a special case, the subalgebra t of s consists of all elements α for which the norm

$$\|\alpha\|_\infty = \sup_n |\alpha_n| \tag{1}$$

is finite, then the underlying vector space of s is the Banach space l_∞. Since $\|\alpha \beta\|_\infty \leqslant \|\alpha\|_\infty \|\beta\|_\infty$, $\alpha, \beta \in l_\infty$, l_∞ is also a commutative B-algebra. Hereafter, we use common symbols for the commutative B-algebras and the corresponding complex Banach spaces.

Lemma 2. *Let* $t \subset s$ *be a B-algebra such that* $\delta_n \in t$ *for all n. Then t is a subset of* l_∞.

Proof. Let $\| \cdot \|$ be the norm on t. Since $|\alpha_n| \|\delta_n\| = \|\alpha \delta_n\| \leqslant \|\alpha\| \|\delta_n\|$, it follows immediately from the definition of $\|\alpha_\infty\|$ that $\|\alpha\|_\infty \leqslant \|\alpha\|$, hence that t is in l_∞.

Definition 3. *A commutative B-algebra* $t \subset s$ *with norm* $\| \cdot \|$ *on t is called a π-ring if and only if* $e, \delta_n \in t$, $n = 1, 2, ...$ *and there is a constant* $K > 0$ *such that* $\sup_n \|\chi_n \alpha\| \leqslant K \|\alpha\|$ *for every α in t, where* $\chi_n \in t$ *is defined by* $\chi_n = \sum_{i \leqslant n} \delta_i$.

The commutative B-algebras l_∞ and c based on the complex Banach spaces l_∞ and c are π-rings. This is clear since the norms in both l_∞ and c are defined by (1). It is called attention to the fact that every π-ring is a linear subset of l_∞ though its topology is not the same as that of l_∞ in general. Since a π-ring t is a semi-simple commutative B-algebra it follows from a known theorem (I.1.18):

Lemma 4. *The norms by which a subalgebra t of s forms a π-ring are all equivalent.*

Next, let $\bar{t}$ be the closure of a π-ring t in the topology of l_∞. Then the identity mapping of t into the closed linear subspace $\bar{t}$ of l_∞ is continuous by the inequality $\|\alpha\|_\infty \leqslant \|\alpha\|$ derived in the proof of Lemma 2. There are two possibilities: Either $t \neq \bar{t}$ or $t = \bar{t}$ (as sets). In the second case, by (I.1.14) the norm $\| \cdot \|$ is equivalent to the norm $\| \cdot \|_\infty$.

Definition 5. *A π-ring t is said to be of the second category in l_∞ if and only if $t = \bar{t}$. Otherwise t is said to be of the first category.*

Clearly l_∞ and c are π-rings of the second category in l_∞. To show the existence of a π-ring of the first category in l_∞, we need the following lemma.

Lemma 6. *If t is a π-ring of the second category in l_∞, then $c \subset t$.*

Proof. Let $\alpha \in c$ and define $a \in \mathbb{C}$ by $a = \lim_n \alpha_n$. Then $e, \delta_i \in t$, $i = 1, 2, \ldots$, imply that $\beta_n = ae + \sum_{i \leqslant n} (\alpha_i - a)\delta_i \in t$. Because β_n converges to α in the topology of l_∞, α is in $\bar{t}$. But $\bar{t} = t$ since t is of the second category in l_∞, and we obtain $\alpha \in t$.

Lemma 7. *Let $w \subset s$ be the set of all α in s for which $\sum_{i=1}^{\infty} |\alpha_i - \alpha_{i+1}| < \infty$. Then*

(i) *w is properly contained in c,*
(ii) *the function $\| \ \|_w : w \to \mathbb{R}$ defined by $\|\alpha\|_w = \sup_i |\alpha_i| + \sum_{i=1}^{\infty} |\alpha_i - \alpha_{i+1}|$ is a norm and*
(iii) *w is a π-ring of the first category in l_∞.*

Proof. (i) Let $\alpha \in w$. From $|\alpha_p - \alpha_q| = \left| \sum_{i=p}^{q-1} (\alpha_i - \alpha_{i+1}) \right| \leqslant \sum_{i=p}^{q-1} |\alpha_i - \alpha_{i+1}|$, $p < q$ and the convergence of $\sum_{i=1}^{\infty} |\alpha_i - \alpha_{i+1}|$ it follows that α_i is convergent, hence that α is in c. Thus $w \subset c$. However, since $\{1, -\frac{1}{2}, \frac{1}{3}, -\frac{1}{4}, \ldots\}$ is an element of c, but not of w, it is evident that w is properly contained in c.

(ii). Obviously $\|\alpha\|_w < \infty$ for each α in w and the function $\| \ \|_w : w \to \mathbb{R}$ is a norm on w.

(iii). We first prove that w is complete. Let $\{\beta_p\}$ be a Cauchy sequence in w. Then we have for every $\varepsilon > 0$ an n such that $\|\beta_p - \beta_q\|_w < \varepsilon$, $p, q \geqslant n$. From the definition of norm we thus obtain $|\beta_{pi} - \beta_{qi}| < \varepsilon$, $p, q \geqslant n$. Hence there is a β_0 in s such that $\lim_p \beta_{pi} = \beta_{0i}$, $i = 1, 2, \ldots$. Moreover, for any m there is a $q \geqslant n$ such that $\sum_{i \leqslant m} |\beta_{qi} - \beta_{0i} - (\beta_{q,i+1} - \beta_{0,i+1})| < \varepsilon$. Since then

$$\|\beta_p - \beta_0\|_w = \sup_i |\beta_{pi} - \beta_i| + \sup_m \sum_{i \leqslant m} |\beta_{pi} - \beta_{0i} - (\beta_{p,i+1} - \beta_{0,i+1})|$$

$$\leqslant \varepsilon + \sup_{q \geqslant n} \|\beta_p - \beta_q\| + \varepsilon \leqslant 3\varepsilon, \qquad p \geqslant n,$$

β_0 is in w and $\lim_p \beta_p = \beta_0$.

Next we show that w is a (commutative) B-algebra. As a direct result of the definition of w we see that the identity e is contained in w and that $\|e\|_w = 1$. Let $\alpha, \beta \in w$. Since then $\alpha, \beta \in c$ and

$$\sum_{i=1}^{\infty} |\alpha_i \beta_i - \alpha_{i+1} \beta_{i+1}| \leqslant \sum_{i=1}^{\infty} [|(\alpha_i - \alpha_{i+1})\beta_i| + |(\beta_i - \beta_{i+1})\alpha_{i+1}|]$$

$$\leqslant \sup_j |\beta_j| \cdot \sum_{i=1}^{\infty} |\alpha_i - \alpha_{i+1}| + \sup_j |\alpha_j| \cdot \sum_{i=1}^{\infty} |\beta_i - \beta_{i+1}|,$$

it follows that $\alpha\beta \in w$. We verify that $\|\alpha\beta\|_w \leqslant \|\alpha\|_w \|\beta\|_w$:

$$\|\alpha\beta\|_w = \sup_i |\alpha_i \beta_i| + \sum_{i=1}^{\infty} |\alpha_i \beta_i - \alpha_{i+1}\beta_{i+1}|$$

$$\leqslant \sup_i |\alpha_i| \sup_j |\beta_j| + \sup_j |\beta_j| \sum_{i=1}^{\infty} |\alpha_i - \alpha_{i+1}| + \sup_j |\alpha_j| \sum_{i=1}^{\infty} |\beta_i - \beta_{i+1}|$$

$$\leqslant \|\alpha\|_w \|\beta\|_w \,.$$

Now, w is a π-ring, because $\delta_n \in w$, $n = 1, 2, \dots$ and since

$$\|\chi_n \alpha\|_w = \sup_{i \leqslant n} |\alpha_i| + \sum_{i=1}^{n-1} |\alpha_i - \alpha_{i+1}| + |\alpha_n|$$

$$\leqslant 2 \sup_i |\alpha_i| + \sum_{i=1}^{\infty} |\alpha_i - \alpha_{i+1}| \leqslant 2 \|\alpha\|_w \,.$$

Finally, w cannot be of the second category in l_∞, since by Lemma 6 this would imply $w = c$ which contradicts (i). This completes the proof of the lemma.

3. Proper π-Rings of Schauder Decompositions

In this section we discuss endomorphisms $A(\alpha)$ of a Banach space X which are related somehow to an element α of the commutative algebra s. Let $\{M_i\}$ be a Schauder decomposition of X with associated sequence of projections $P_1, P_2, \dots$ of X onto the nontrivial subspaces $M_1, M_2, \dots$ respectively. By r we define the following subset of s:

$$r = \left\{ \alpha \,\middle|\, \alpha \in s, \quad \lim_n \sum_{i \leqslant n} \alpha_i P_i x \quad \text{exists for all} \quad x \in X \right\}.$$

To each $a \in r$ there corresponds a linear transformation $A(\alpha): X \to X$, given by

$$A(\alpha)x = \lim_n \sum_{i \leqslant n} \alpha_i P_i x. \tag{1}$$

We have

Lemma 1. $A(\alpha)$ *is bounded for every* $\alpha \in r$.

Proof. Let $\alpha \in r$. Since $A(\chi_n \alpha)$ is represented by a finite sum and all of the P_i's are bounded, $A(\chi_n \alpha)$ is an endomorphism of X. The convergence of (1) implies that $\sup_n \|A(\chi_n \alpha)x\| < \infty$, $x \in X$. As a consequence of the Banach-Steinhaus theorem (I.2.14) it follows $\|A(\alpha)\| < \infty$.

Next, we define the subset $\Pi = \{A(\alpha) | \alpha \in r\}$ of $B(X)$. The following lemma shows how Π is related to r.

Lemma 2. *The linear transformation* $A : r \to \Pi$ *is one-to-one.*

Proof. Let $A(\alpha) = 0$. Due to the orthogonality property $P_i P_j = \delta_{ij} P_j$ we then have $0 = A(\alpha) P_i x = \alpha_i P_i x$ for all $x \in X$ and all i. Since the subspaces M_i are nontrivial, each P_i must be different from zero. Thus $\alpha_i = 0$ for all i, i.e. $\alpha = 0$ and this shows that A is one-to-one.

Lemma 3. r *is a subset of the space* l_∞.

Proof. Let $\alpha \in r$. Again from $A(\alpha) P_i x = \alpha_i P_i x, x \in X$ we obtain $|\alpha_i| \|P_i x\| \leqslant \|A(\alpha)\| \|P_i x\|$, hence $\|\alpha\|_\infty = \sup_i |\alpha_i| \leqslant \|A(\alpha)\| < \infty$.

Theorem 4. Π *is a semi-simple commutative B-algebra. There exists a norm on* r *such that* r *is a* π-*ring and that* A *is a proper isometric isomorphism of* r *onto* Π.

Proof. Using the orthogonality relation $P_i P_j = \delta_{ij} P_j$ one gets for $A(\alpha)$, $A(\beta) \in \Pi$, $A(\alpha) A(\beta) x = \lim_n \sum_{i \leqslant n} \alpha_i P_i A(\beta) x = \lim_n \sum_{i \leqslant n} \alpha_i \beta_i P_i x = A(\alpha \beta) x$, $x \in X$ and thus $A(\alpha \beta) \in \Pi$. Since $A(e) x = \lim_n \sum_{i \leqslant n} P_i x = x$, $x \in X$, one has $A(e) = I$. Therefore A is a proper (algebraic) isomorphism of r onto π.

Moreover, Π is a Banach space: Let $\{\alpha_m\}$ be a sequence in r such that $A(\alpha_m)$ converges in the uniform operator topology of $B(X)$, say to $A_0 \in B(X)$. Since $A(\alpha_m) P_i x = \alpha_{mi} P_i x$ for each i and each $x \in X$, we have an $\alpha_0 \in s$ such that $\lim_m \alpha_{mi} = \alpha_{0i}$. Moreover, $\lim_m A(\alpha_m) P_i x = A_0 P_i x$. From $\sum_{i \leqslant n} \alpha_{0i} P_i x = \lim_m \sum_{i \leqslant n} \alpha_{mi} P_i x = \lim_m A(\alpha_m) \sum_{i \leqslant n} P_i x = A_0 \sum_{i \leqslant n} P_i x$ and the boundedness of A_0 we obtain that $A_0 = A(\alpha_0) \in \Pi$. This shows that Π is a closed subset of $B(X)$, and since $B(X)$ is complete, Π is also complete.

Through the mapping $\|\cdot\| : r \to \mathbb{R}$, given by $\|\alpha\| = \|A(\alpha)\|$, we now introduce a norm on r and finally show that r is a π-ring. Corollary VII.2.8 implies the existence of a constant $K \geqslant 1$ such that

$$\|A(\chi_n \alpha)\| = \|A(\alpha) A(\chi_n)\| = \left\| A(\alpha) \sum_{i \leqslant n} P_i \right\| \leqslant K \|A(\alpha)\|$$

for all $\alpha \in r$. Hence $\|\chi_n \alpha\| = \|A(\chi_n \alpha)\| \leqslant K \|A(\alpha)\| = K \|\alpha\|$, and for $\alpha, \beta \in r$ we have $\|\alpha \beta\| = \|A(\alpha \beta)\| = \|A(\alpha) A(\beta)\| \leqslant \|A(\alpha)\| \, \|A(\beta)\| = \|\alpha\| \, \|\beta\|$. The completeness of r follows from the fact that A is an isometric isomorphism of r onto Π. Because of these properties, and since $e, \delta_n \in r$, $n = 1, 2, \ldots, r$ is a π-ring. Finally, the semi-simplicity of Π is a consequence of the semi-simplicity of r, where the semi-simplicity of r follows from Lemma 2.1.

Definition 5. r is said to be the proper π-ring of the Schauder decomposition $\{M_i\}$ of X.

The next theorem shows that the notions of strong and weak proper π-rings of Schauder decompositions of X (corresponding to strong and weak convergence of the series expansion which defines the operator $A(\alpha)$) are the same.

Theorem 6. Let $\{M_i, P_i\}$ be a Schauder decomposition of X. If for some α in s the series $\sum\limits_{i=1}^{\infty} \alpha_i P_i x$ is weakly convergent for each $x \in X$, then it is convergent in the strong topology for X for all $x \in X$.

Proof. Let r be the proper π-ring of $\{M_i\}$. Obviously, $\chi_n \alpha \in r$ for every finite n. By hypothesis, $x^*[A(\chi_n \alpha) x]$ converges for each x in X and each x^* in X^*. Thus $\sup\limits_n |x^*(A(\chi_n \alpha) x)| < \infty$, $x \in X$, $x^* \in X^*$ which, as a consequence of the uniform boundedness principle (I.3.14), implies $\sup\limits_n \|A(\chi_n \alpha)\| \leqslant M$, where M is some positive constant. Since for every x in X and every $\varepsilon > 0$ we have an index p such that $\|x - x_p\| < \varepsilon / 2M$, where $x_p = \sum\limits_{i \leqslant p} P_i x$, we have

$$\left\| \sum_{i=m}^{n} \alpha_i P_i x \right\| \leqslant \left\| \sum_{i=m}^{n} \alpha_i P_i (x - x_p) \right\| + \left\| \sum_{i=m}^{n} \alpha_i P_i x_p \right\|$$

$$= \left\| [A(\chi_n \alpha) - A(\chi_{m-1} \alpha)] (x - x_p) \right\|$$

$$\leqslant 2M \|x - x_p\| < \varepsilon,$$

for every $n \geqslant m > p$. Thus $\left\{ \sum\limits_{i \leqslant n} \alpha_i P_i x \right\}$ is a Cauchy sequence and since X is complete, it is shown that the series converges strongly in X.

Theorem 7. A sequence of closed subspaces $\{M_i\}$ in X is a Schauder decomposition of X if and only if the following conditions are satisfied:

(i) There is an isometric isomorphism A from a π-ring r into $B(X)$,

(ii) for each i, $A(\delta_i)$ is a projection of X onto M_i and

(iii) $X = \overline{\text{sp}} \bigcup\limits_{i=1}^{\infty} M_i$.

Proof. Necessity. (i) is a consequence of Theorem 4, (ii) follows from the definition (1) of A which yields $A(\delta_i) = P_i$, and (iii) is a necessary condition for $\{M_i\}$ to be a Schauder decomposition of X.

Sufficiency. If (i) is satisfied, then since A is a homomorphism, $A(\delta_i) A(\delta_j) = A(\delta_i \delta_j) = \delta_{ij} A(\delta_j)$. Defining $P_i = A(\delta_i)$, which, by (ii) is a projection of X onto M_i, one obtains $P_i P_j = \delta_{ij} P_j$. Since A is an isometric isomorphism we have $\|\alpha\| = \|A(\alpha)\|$ on r, where $\|\alpha\|$ is the norm of $\alpha \in r$. Taking now $\alpha = \chi_n$ yields $\left\| \sum_{i \leqslant n} P_i x \right\| = \|A(\chi_n) x\| \leqslant \|\chi_n\| \|x\|$. Since r is a π-ring there exists a constant $K \geqslant 1$ such that $\|\chi_n\| \leqslant K \|e\| = K$. This implies $\left\| \sum_{i \leqslant n} P_i x \right\| \leqslant K \|x\|$ and by Corollary VII.2.8, $\{M_i\}$ is a Schauder decomposition of X with associated sequence of projections $\{P_i\}$.

Corollary 8. *The π-ring r of the above theorem is the proper π-ring of $\{M_i\}$.*

Proof. Since for $\alpha \in r$, $\quad \sum_{i \leqslant n} \alpha_i P_i x = \sum_{i \leqslant n} \alpha_i A(\delta_i) x = \sum_{i \leqslant n} A(\alpha \delta_i) x$ $= A(\alpha) \sum_{i \leqslant n} P_i x$ and since $\{M_i\}$ is a Schauder decomposition of X, $\lim_{n} \sum_{i \leqslant n} \alpha_i P_i x$ exists for each $x \in X$. Hence $A(\alpha) x = \lim_{n} \sum_{i \leqslant n} \alpha_i P_i x$, showing that r is the proper π-ring of $\{M_i\}$.

Corollary 9. *Theorem 7 holds if the condition* (i) *is replaced by:* (i') *There is an isomorphism A from a π-ring r onto a B-algebra $\Pi \subset B(X)$.*

Proof. Again, by Theorem 4, (i') is a necessary condition. Since now A is only an isomorphism, in the sufficiency proof one has to show that A is homeomorphic: Since a π-ring is semi-simple and the semi-simplicity of r implies the semi-simplicity of $\Pi = A(r)$, by a known theorem (I.1.19), A is a homeomorphism. Hence there exists a constant $M > 0$ such that $\sup \{\|A(\alpha)\| \,\big|\, \|\alpha\| \leqslant 1, \alpha \in r\} \leqslant M$. We now obtain $\left\| \sum_{i \leqslant n} P_i x \right\| \leqslant KM \|x\|$ and the rest of the proof is the same as that for Theorem 7.

In the following we investigate the problem of the extension of a topological isomorphism from a proper π-ring r of a Schauder decomposition of X, onto a B-algebra $\Pi \subset B(X)$.

Definition 10. *A topological isomorphism from r into $B(X)$ is π-maximal whenever it cannot be extended to a topological isomorphism from r' into $B(X)$, where r' is a π-ring properly containing r.*

Theorem 11. *Let r be the proper π-ring of a Schauder decomposition of X. Then the isometric isomorphism A of r into $B(X)$ is π-maximal.*

Proof. If A is not π-maximal, then there exists a topological isomorphism A' of a π-ring r' which contains r properly, onto a B-algebra $\Pi' \supset \Pi$. Let then α be an element of r' which is not in r. Because $\chi_n \alpha \in r$, $n = 1, 2, \ldots,$

$$A'(\chi_n \alpha) x = \sum_{i \leqslant n} \alpha_i P_i x \tag{2}$$

holds for every x in X. Let K' be the constant of Definition 2.3 which corresponds to r'. Then

$$\left\| \sum_{i \leqslant n} \alpha_i P_i x \right\| \leqslant \| A'(\chi_n \alpha) \| \, \| x \| \leqslant \| A' \| \, \| \chi_n \alpha \|_{r'} \, \| x \| \leqslant K' \| A' \| \, \| \alpha \|_{r'} \, \| x \| .$$

Since $x = \lim\limits_n \sum\limits_{i \leqslant n} P_i x$ we have for each x in X and every $\varepsilon > 0$ an index n such that for $p, q \geqslant n$, $\left\| \sum\limits_{j=p}^{q} P_j x \right\| < \varepsilon$. Thus

$$\left\| \sum_{i=p}^{q} \alpha_i P_i x \right\| = \left\| \sum_{i \leqslant q} \alpha_i P_i \sum_{j=p}^{q} P_j x \right\| < K' \| A' \| \, \| \alpha \|_{r'} \, \varepsilon$$

and since X is complete, the right hand side of (2) is convergent. Therefore α is in r and this contradiction leads to the desired result that A is π-maximal.

The concept of a proper π-ring can be extended in an obvious way to a weak* Schauder basis $\{x_i^*, x_i^{**}\}$ of X^*. Let r^* be the subset of all α in s such that $\sum\limits_{i=1}^{\infty} \alpha_i x_i^{**}(x^*) x_i^*$ exists in the weak* topology for all x^* in X^*. For $\alpha \in r^*$ we then define the mapping $A^+(\alpha): X^* \to X^*$ by

$$A^+(\alpha) x^*(x) = \sum_{i=1}^{\infty} \alpha_i x_i^{**}(x^*) x_i^*(x), \qquad x \in X, \qquad x^* \in X^*.$$

Since $\{x_i\}$ is a basis for X with biorthogonal sequence $\{x_i^*\}$ if and only if $\{x_i^*\}$ is a weak* Schauder basis for X^* with biorthogonal sequence $\{x_i^{**}\}$ such that $x_i^{**}(x^*) = x^*(x_i)$ for all $x^* \in X^*$ (III.2.7), we have

$$A^+(\alpha) x^*(x) = \sum_{i=1}^{\infty} \alpha_i x_i^*(x) x^*(x_i), \qquad x \in X, \qquad x^* \in X^*. \tag{3}$$

Next, let r be the proper π-ring of $\{x_i\}$ and A the corresponding isometric isomorphism of r into $B(X)$. From (3) it is apparent that $A^+(\chi_n \alpha)$ is the adjoint of $A(\chi_n \alpha)$ for all n and, as a consequence of the uniform boundedness principle (I.3.14), $\sup\limits_n \| A(\chi_n \alpha) \| < \infty$. Since $\lim\limits_n A(\chi_n \alpha) x$ exists in X for each x in the total set $\{x_i\}$, it follows from the Banach-Steinhaus theorem that $A(\alpha) x$ exists for all x in X. Hence $r^* \subset r$ and by (1) and (3),

$$x^*(A(\alpha) x) = A^+(\alpha) x^*(x), \qquad x \in X, \qquad x^* \in X^*,$$

for all α in r. This immediately shows that $A^+(\alpha) = A^*(\alpha)$, the adjoint of $A(\alpha)$, and that $r^* = r$ (as sets). Defining analogously the norm on r^* by $\|\alpha\| = \|A^*(\alpha)\|$ and using the fact that $B(X)$ is isometrically isomorphic to $B(X^*)$ (I.3.23), it is clear that r^* is a π-ring, the proper π-ring of the weak* Schauder basis $\{x_i^*\}$ for X^*, and one gets the following result:

Theorem 12. *A weak* Schauder basis for X^* and its corresponding basis for X have the same proper π-ring.*

4. Minimal Schauder Decompositions

Let Γ be the set of all Schauder decompositions of Banach spaces. Then it may happen that the proper π-ring r_γ of an element γ in Γ is contained (as a set) in the proper π-ring $r_{\gamma'}$ for all γ' in Γ. In this case we write

Definition 1. *A Schauder decomposition γ in Γ is said to be minimal if $r_\gamma \subset r_{\gamma'}$ for every γ' in Γ; then r_γ is called the proper minimal π-ring of γ.*

To show the existence of a minimal Schauder decomposition we use the following lemmas and remember that every basis for a Banach space X automatically generates, by its basis elements, a Schauder decomposition of X.

Lemma 2. *The subalgebra w of s (cf. Lemma 2.7) is included in the proper π-ring r_γ for each γ in Γ.*

Proof. Let X be the corresponding Banach space of an arbitrary $\gamma = \{P_i(X), P_i\}$ in Γ with proper π-ring r_γ. Now, for any α in w we have by Abel's formula (which is to be verified by an elementary calculation)

$$\sum_{i \leq n} \alpha_i P_i = \sum_{i \leq n} (\alpha_i - \alpha_{i+1}) \sum_{j \leq i} P_j + \alpha_{n+1} \sum_{i \leq n} P_i. \tag{1}$$

By Corollary VII.2.8 there exists a positive constant K such that $\sup_n \left\| \sum_{i \leq n} P_i x \right\| \leq K \|x\|$ for every x in X. Therefore

$$\left\| \sum_{i \leq n} (\alpha_i - \alpha_{i+1}) \sum_{j \leq i} P_j x \right\| \leq \sum_{i \leq n} |\alpha_i - \alpha_{i+1}| \left\| \sum_{j \leq i} P_j x \right\|$$

$$\leq K \|x\| \sum_{i \leq n} |\alpha_i - \alpha_{i+1}|. \tag{2}$$

Combining (1) and (2) yields

$$\|A(\chi_n \alpha) x\| = \left\| \sum_{i \leq n} \alpha_i P_i x \right\| \leq K \|x\| \left(|\alpha_{n+1}| + \sum_{i \leq n} |\alpha_i - \alpha_{i+1}| \right) \leq K \|x\| \|\alpha\|_w.$$

Since $\{P_i(X), P_i\}$ is a Schauder decomposition of X we have for each x in X and every $\varepsilon > 0$ an index n such that for $p, q \geqslant n$, $\left\| \sum\limits_{i=p}^{q} P_i x \right\| < \varepsilon$. Therefore,

$$\left\| \sum_{i=p}^{q} a_i P_i x \right\| = \left\| \sum_{i \leqslant q} \alpha_i P_i \sum_{j=p}^{q} P_j x \right\| < K \cdot \|\alpha\|_w \varepsilon$$

and because X is complete, $\sum\limits_{i \leqslant n} \alpha_i P_i x$ converges in X. Moreover, $\|A(\alpha)\| \leqslant K \|\alpha\|_w$ for every α in w. This shows that w, as a set, is contained in r.

We now introduce the ideal w_0 of w, defined by

$$w_0 = \{\alpha \,|\, \alpha \in w, \ \lim_n \alpha_n = 0\} \,.$$

In other words we have $w_0 = w \cap c_0$. Then we show the following

Lemma 3. *The sequence $\{\delta_i\}$ is a basis for the Banach space w_0.*

Proof. We recall that the norm in w_0 is given by $\|\alpha\|_w = \sup\limits_i |\alpha_i| + \sum\limits_{i=1}^{\infty} |\alpha_i - \alpha_{i+1}|$, $\alpha \in w_0$. For any α in w_0, we have $\chi_n \alpha \in w_0$ and

$$\|\alpha - \chi_n \alpha\|_w \leqslant 2 \sup_{i \leqslant n} |\alpha_{n+i}| + \sum_{i=1}^{\infty} |\alpha_{n+i} - \alpha_{n+i+1}| \,.$$

Since $\alpha \in w_0 \subset c_0$, the right hand side of the above inequality tends to zero with increasing n. Thus $\overline{\mathrm{sp}} \{\delta_i\} = w_0$. On the other hand it is clear that for $m \geqslant n$ and any β in s,

$$\left\| \sum_{i \leqslant n} \beta_i \delta_i \right\|_w = \|\chi_n \beta\|_w \leqslant \|\chi_m \beta\|_w = \left\| \sum_{i \leqslant m} \beta_i \delta_i \right\|_w \,.$$

Hence, by Theorem IV.1.5, $\{\delta_i\}$ is a basis for w_0.

Lemma 4. *w is the proper π-ring of the basis $\{\delta_i\}$ for w_0.*

Proof. We know from the preceding lemma that $\{\delta_i\}$ is a basis for w_0 and we denote the corresponding proper π-ring by r. By Lemma 2, $w \subset r$. Conversely, we can show that $r \subset w$: Let $\alpha \in r$. From the Lemmas 3 and 2.7 we then infer $\sup\limits_i |\alpha_i| + \sum\limits_{i=1}^{n} |\alpha_i - \alpha_{i+1}| = \|\chi_n \alpha\|_w + |\alpha_n - \alpha_{n+1}|$

$$-|\alpha_n| \leqslant \left\| \sum_{i \leqslant n} \alpha_i \delta_i \right\|_w + |\alpha_{n+1}| = \|A(\alpha) \chi_n\|_w + \|\alpha \delta_{n+1}\| / \|\delta_{n+1}\| \leqslant \|A(\alpha)\| \ \|\chi_n\|_w$$

$+ \|\alpha\| = 3 \|\alpha\|$, where $\| \ \|_w$ is the norm used in the space w_0 and $\| \ \|$ is the norm used in r. From the definition of the norm in w it is now clear that $\|\alpha\|_w \leqslant 3 \|\alpha\|$, hence that $\alpha \in w$. As desired, this implies $r = w$.

Theorem 5. *The basis $\{\delta_i\}$ for w_0 is minimal, its proper π-ring is w.*

This theorem is an immediate consequence of the preceding three lemmas.

Theorem 6. *A sequence of closed linear subspaces $\{M_i\}$ in a Banach space X is a Schauder decomposition of X if and only if the following conditions are satisfied:*
(i) *There exists a continuous isomorphism A from the π-ring w into $B(X)$,*
(ii) *for each i, $A(\delta_i)$ is a projection of X on M_i and*

(iii) $X = \overline{\mathrm{sp}} \bigcup_{i=1}^{\infty} M_i$.

Proof. In view of the inequality $\|A(\alpha)\| \leqslant K \|\alpha\|_w$, $\alpha \in w$, derived in the proof of Lemma 2, the necessity is an immediate consequence of Theorem 3.7 and Lemma 2. The sufficiency proof is analogous to that of Theorem 3.7, where instead of r one only has to take the π-ring w.

5. Banach Algebras and Unconditional Bases

Definition 1. *A basis for a Banach space X is of the first or second category according to its proper π-ring is of the first or second category.*

Theorem 2. *A basis for X is of the second category if and only if there is a positive constant M such that $\sup_n \|A(\chi_n \alpha)\| \leqslant M \|\alpha\|_\infty$ for all α in l_∞.*

Proof. To show the necessity, let the basis be of the second category. Then there exists a constant $M > 0$ such that $\|\alpha\| \leqslant M \|\alpha\|_\infty$ for each α in the proper π-ring r of the basis. This is clear by (I.2.6) since $\|\alpha\|_\infty \leqslant \|\alpha\|$ for each α in r (VIII.2.2) and since the identity maps r in a one-to-one manner onto the Banach space $\bar{r}$ (with the topology of l_∞). Evidently $\chi_n \alpha$ is in r for each α in l_∞. Hence $\|A(\chi_n \alpha)\| = \|\chi_n \alpha\| \leqslant M \|\chi_n \alpha\|_\infty \leqslant M \|\alpha\|_\infty$ and the assertion follows. On the other hand, $\sup_n \|A(\chi_n \alpha)\| \leqslant M \|\alpha\|_\infty$ for all α in r is a sufficient condition for the basis to be of the second category, because then $\|A(\alpha)x\| \leqslant \sup_n \|A(\chi_n \alpha)x\| \leqslant M \|\alpha\|_\infty \|x\|$ for every x in X. Consequently, $\|\alpha\| = \|A(\alpha)\| \leqslant M \|\alpha\|_\infty$, showing that the convergence in l_∞ implies the convergence in r. Thus, using again (VIII.2.2) $r = \bar{r}$ and this concludes the proof of the theorem.

Theorem 3. *A basis $\{x_i, x_i^*\}$ for X is of the second category if and only if it is unconditional.*

Proof. Let $\{x_i\}$ be unconditional. $A(\alpha)x$ exists in X for each α in l_∞ and each x in X, because unconditional convergence of a series, by Theorem II.1.3, implies bounded multiplier convergence of the series. Hence the proper π-ring of $\{x_i\}$ contains l_∞ and since $r \subset l_\infty$ we have, as a set, $r=l_\infty$. We now proceed to define the transformations $T_n:l_\infty \to B(X)$ by $T_n(\alpha)=A(\chi_n\alpha), \alpha\in l_\infty, n=1,2,\ldots$. Since for fixed α in l_∞, $\sup_n \|T_n(\alpha)x\|$ $=\sup_n \|A(\chi_n\alpha)x\| < \infty$, $x\in X$, we have by the principle of uniform boundedness (I.3.14), $\sup_n \|T_n(\alpha)\| < \infty, \alpha\in l_\infty$. Evidently each of the transformations T_n is bounded. Applying now the same principle again it then follows that $\sup_n \sup \{\|A(\chi_n\alpha)\| \mid \|\alpha\|_\infty \leqslant 1\} = \sup_n \|T_n\| < \infty$ so that by Theorem 2, $\{x_i\}$ is of the second category.

Conversely, if $\{x_i\}$ is of the second category, then $\|A(\chi_n\alpha)\| \leqslant M\|\alpha\|_\infty$ for all α in l_∞ and all n. Let $\{n_i\}$ be an arbitrary increasing sequence of integers and let α in l_∞ be such that $\alpha_j=1$ for j in $\{n_i\}$ and $\alpha_j=0$ else. For every $\varepsilon>0$ we have an index p such that $\|x-x_p\| < \varepsilon/(2M)$, where $x_p = \sum_{i\leqslant p} x_i^*(x)x_i$ and where x is a fixed but arbitrary element of X. We show that $\left\{\sum_{i\leqslant n} \alpha_i x_i^*(x)x_i\right\}$ is a Cauchy sequence: If $n\geqslant m>p$,

then $\left\|\sum_{i=m}^n \alpha_i x_i^*(x)x_i\right\| \leqslant \left\|\sum_{i=m}^n \alpha_i x_i^*(x-x_p)x_i\right\| + \left\|\sum_{i=m}^n \alpha_i x_i^*(x_p)x_i\right\|$ $=\|[A(\chi_n\alpha)-A(\chi_{m-1}\alpha)](x-x_p)\| \leqslant 2M\|x-x_p\| < \varepsilon$. Since $\sum_{i\leqslant n} \alpha_i x_i^*(x)x_i$ $=\sum_{n_i\leqslant n} x^*(x)x_i$, and $\{n_i\}$ was supposed to be arbitrary, the basic series $\sum_{i=1}^\infty x_i^*(x)x_i$ is subseries convergent, hence unconditionally convergent (II.1.3) and the theorem is proved.

Corollary 4. *An element α of s is in the proper π-ring of an unconditional basis if and only if it is in l_∞.*

Corollary 5. $\{x_i\}$ *is an unconditional basis for X if and only if the following conditions are satisfied:*
  (i) *There exists a topological isomorphism A of l_∞ into $B(X)$,*
  (ii) *for each i, $A(\delta_i)$ is a projection of X with one-dimensional range spanned by x_i and*
(iii) $\overline{sp}\{x_i\}=X$.

Proof. The necessity is a consequence of Theorem 3.7, the preceding corollary and Theorem I.1.14. Conversely, with $\|A\| < \infty$ and in just the same way as in the sufficiency proof of Theorem 3.7 it follows that $\{x_i\}$ is a basis for X. Since the π-ring is l_∞, $\{x_i\}$ is, by definition, of the second category and thus, by Theorem 3, unconditional. This finishes the proof of the corollary.

Theorem 6. *Let $\{x_i, x_i^*\}$ be an unconditional basis for X and let r be its proper π-ring. Then the spectrum of $A(\alpha)$, $\alpha\in r$, is the closure of the set*

of eigenvalues $\{\alpha_1, \alpha_2, ...\}$ *of* A *in* $\mathbb{C}$ *and the resolvent* $R(\lambda, A(\alpha))$ *of* $A(\alpha)$
is given by $R(\lambda, A(\alpha)) = A((\lambda e - \alpha)^{-1})$.

Proof. Since $A(\alpha) x_i = \lim\limits_{n} \sum\limits_{j \leqslant n} \alpha_j x_j^*(x_i) x_j = \alpha_i x_i$, each value a_i in $\mathbb{C}$ is an

eigenvalue of $A(\alpha)$ with eigenfunction x_i. On the other hand, if λ is not
in the closure of the set $\{\alpha_1, \alpha_2, ...\}$ in $\mathbb{C}$, then we have $\inf\limits_{i} |\lambda - \alpha_i| > 0$.

Now, from $(\lambda I - A(\alpha)) x = \lim\limits_{n} \sum\limits_{i \leqslant n} (\lambda - \alpha_i) x_i^*(x) x_i, \ x \in X$, through multipli-

cation of both sides by $(\lambda - \alpha_j)^{-1} x_j^*$, one obtains $(\lambda - \alpha_j)^{-1} x_j^* (\lambda I - A(\alpha)) x$
$= x_j^*(x)$ for all j. Since $x = \lim\limits_{n} \sum\limits_{i \leqslant n} x_i^*(x) x_i$, the assumption $(\lambda I - A(\alpha)) x = 0$

then implies $x = 0$ for every x in X. Thus $\lambda I - A(\alpha)$ is one-to-one and this
property shows that $(\lambda I - A(\alpha))^{-1}$ exists. Because $(\lambda e - \alpha)^{-1}$ is in l_∞,
Corollary 4 implies that $(\lambda e - \alpha)^{-1}$ is in r. Therefore, $A((\lambda e - \alpha)^{-1})$ is
bounded on X. Hence $A((\lambda e - \alpha)^{-1}) x = (\lambda I - A(\alpha))^{-1} (\lambda I - A(\alpha))$
$A((\lambda e - \alpha)^{-1}) x = (\lambda I - A(\alpha))^{-1} \lim\limits_{n} \sum\limits_{i \leqslant n} (\lambda - \alpha_i)^{-1} x_i^*(x)(\lambda - \alpha_i) x_i = (\lambda I - A(\alpha))^{-1} x$

for all x in X, so that $R(\lambda, A(\alpha)) = (\lambda I - A(\alpha))^{-1} = A((\lambda e - \alpha)^{-1})$. Finally,
since the spectrum of $A(\alpha)$ is closed in $\mathbb{C}$ (I.2.20), and since $R(\lambda, A(\alpha))$
is bounded for each complex λ which is not in the closure of $\{\alpha_1, \alpha_2, ...\}$
in $\mathbb{C}$, the proof of the theorem is finished.

For the special value $\lambda = 0$ it follows

Corollary 7. *Under the conditions stated in the foregoing theorem,*
whenever α *is in* r, $A(\alpha)$ *has a bounded inverse if and only if* $\inf\limits_{i} |\alpha_i| > 0$.
If this is the case, $A(\alpha)^{-1} = A(\alpha^{-1})$.

The topological isomorphism A may be used to investigate relations
between the unconditional bases for X and the orthogonal bases for X.
The last concept is explained in the following definition.

Definition 8. *A basis* $\{x_i\}$ *for* X *is orthogonal if and only if the ine-*
quality

$$\left\| \sum_{i \in \mu} \alpha_i x_i \right\| \leqslant \left\| \sum_{i \in \mu \cup \nu} \alpha_i x_i \right\|$$

is satisfied for all sequences α *in* s *and all disjoint subsets* μ *and* ν *of* Σ,
where Σ *is the set of all finite subsets of the set of positive integers.*

Theorem 9. (SINGER) *Each unconditional basis for* X *is an orthogonal*
basis for a Banach space X', *where* X' *is the same linear space as* X,
supplied with an equivalent norm.

Proof. Let $\{x_i\}$ be an unconditional basis for X. We then define a
new norm on the linear space X by

$$\|x\|' = \sup\{\|A(\alpha)x\| \mid \|\alpha\|_\infty \leqslant 1, \alpha \in l_\infty\}.$$

This yields

$$\|x\| = \|A(e)x\| \leqslant \sup\{\|A(\alpha)x\| \mid \|\alpha\|_\infty \leqslant 1\} = \|x\|'.$$

On the other hand, as we have shown in the proof of Theorem 2, there is a constant $M > 0$ such that

$$\|x\|' \leqslant \sup\{M\|\alpha\|_\infty \|x\| \mid \|\alpha\|_\infty \leqslant 1\} = M\|x\|.$$

This shows that the primed and the unprimed norms on X are equivalent. Moreover, for $\beta \neq 0$,

$$\begin{aligned}
\|A(\beta)x\|' &= \sup\{\|A(\alpha)A(\beta)x\| \mid \|\alpha\|_\infty \leqslant 1\} \\
&\leqslant \sup\{\|A(\alpha\beta)x\| \mid \|\alpha\|_\infty \leqslant 1\} \\
&= \|\beta\|_\infty \sup\{\|A(\alpha\beta)x\|/\|\beta\|_\infty \mid \|\alpha\|_\infty \leqslant 1\} \\
&\leqslant \|\beta\|_\infty \sup\{\|A(\alpha)x\| \mid \|\alpha\|_\infty \leqslant 1\} = \|\beta\|_\infty \|x\|'.
\end{aligned}$$

Hence the inequality $\|A(\beta)x\|' \leqslant \|\beta\|_\infty \|x\|'$ holds for every β in l_∞ and every x in X. Let $\chi_\mu \in l_\infty$ be the characteristic function of the set $\mu \in \Sigma$. Choosing $\beta = \chi_\mu$ and $x = \sum_{i \in \mu \cup \nu} \alpha_i x_i$, $\nu \in \Sigma$, $\alpha \in s$, we get $\left\|\sum_{i \in \mu} \alpha_i x_i\right\|'$ $= \|A(\chi_\mu)x\|' \leqslant \|\chi_\mu\|_\infty \|x\|' = \|x\|' = \left\|\sum_{i \in \mu \cup \nu} \alpha_i x_i\right\|'$ and this completes the proof of the theorem.

References for Chapter VIII: KADEC [3], SINGER [8] and YAMAZAKI [1–3].

CHAPTER IX

Some Results on Generalized Bases
for Linear Topological Spaces

Having introduced the basis concept for Hilbert and Banach spaces, certain generalizations appear to be at least as important. First of all one discards of almost all requirements used for the definition of a basis. Beginning with the absolute minimum one takes a linear topological space X, does not use the concepts of totalness and countability and avoids all mention of series expansions. Thus, the starting point will be a family $\{x_\lambda\}$ of elements of X and a biorthogonal family $\{f_\lambda\}$ of continuous linear (coefficient) functionals on X. The biorthogonal system $\{x_\lambda, f_\lambda\}$ is said to be maximal if there is no biorthogonal system in which it is properly contained. A biorthogonal system with respect to X is called a generalized basis for X if, in addition, $x \in X$ and $f_\lambda(x) = 0$ for all λ implies $x = 0$. A generalized basis is always a maximal biorthogonal system. If the set of basis elements $\{x_\lambda\}$ of a biorthogonal system $\{x_\lambda, f_\lambda\}$ in X is total in X, then $\{x_\lambda, f_\lambda\}$ is called a dual generalized basis for X. Moreover, if such a basis is also a generalized basis for X, it is called a Markushevich basis for X if the set $\{x_\lambda\}$ is countable, and an extended Markushevich basis for X if $\{x_\lambda\}$ is not countable. Finally, introducing again the concept of a series expansion for elements of X, then a Markushevich basis for X becomes a Schauder basis for X. Hence one has the following hierarchy in terms of increasing generality:

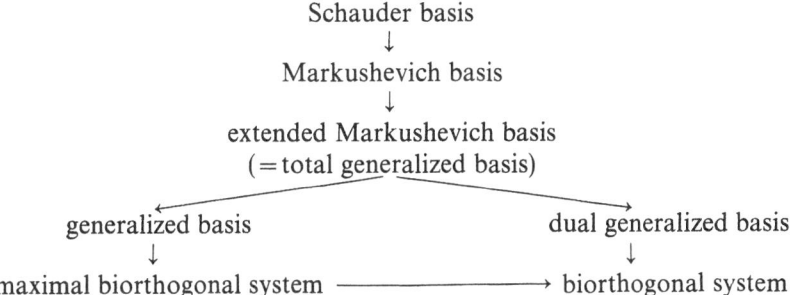

Schauder basis
↓
Markushevich basis
↓
extended Markushevich basis
(= total generalized basis)

generalized basis   dual generalized basis
↓   ↓
maximal biorthogonal system   →   biorthogonal system

Now, let $\{x_\lambda, f_\lambda\}$ be a generalized basis for X. It is shown that $\{x_\lambda\}$ is unique. If X is locally convex, it is also shown that $\{f_\lambda\}$ is unique if and only if $\{x_\lambda\}$ is total in X. On the other hand, the elements $\{x_\lambda\}$ of a dual generalized basis for a locally convex space X are unique if and only if that basis is also a generalized basis for X. Furthermore, if X is also separable and Hausdorff, there exists a dual generalized basis for X, and there is a Markushevich basis for every separable Banach space.

It is worth noting that the basis problem is now solved for separable locally convex linear topological spaces. But it is, as we know, not yet solved for the special case of separable Banach spaces. The solution of the first problem is given in the form of a counter-example (Theorem 1.11): It has been shown that the (Banach) space l_∞^* is weak* separable, but has no basis in the (locally convex) weak* topology. It is also known that there are separable F-spaces with no basis (Corollary 5.3).

After presenting some examples in special spaces we investigate generalized bases for complete linear metric spaces with translation-invariant metric, X and Y, which are similar and show that such bases exist if and only if X and Y are topologically isomorphic. In Banach spaces, bases are similar if and only if they are equivalent. If $X = Y$ and if $\{x_i\}$ is a Schauder basis for the complete linear metric space X with translation-invariant metric, then, as a special case, we obtain the Paley-Wiener theorem which gives sufficient conditions for a sequence in X to be also a Schauder basis for X which is similar to $\{x_i\}$. The final section deals with continuity of the coefficient functionals of bases in complete linear metric spaces and with the connection of bases for the weak and the initial topology respectively in locally convex and in barreled linear topological spaces. The result that every basis for a Banach space is a Schauder basis for that space cannot be generalized to locally convex linear topological spaces, since there exists a basis for the locally convex linear topological space l_1 (endowed with its weak* topology) which is not a Schauder basis (cf. Example III.7.4).

1. Definition and Fundamental Properties of Generalized Bases

Let X be a linear topological space over the field Φ, X^* its conjugate space and let Λ be an index set of arbitrary cardinality.

Definition 1. *The double-family* $\{x_\lambda, f_\lambda\}$ *is a biorthogonal system if* $x_\lambda \in X$ *and* $f_\lambda \in X^*$ *for all* $\lambda \in \Lambda$ *and if* $f_\lambda(x_\mu) = \delta_{\lambda\mu}$, *where* $\delta_{\lambda\mu}$ *is the Kronecker symbol.* $\{x_\lambda, f_\lambda\}$ *is maximal with respect to* X *if there is no biorthogonal system which contains* $\{x_\lambda, f_\lambda\}$ *properly.*

Definition 2. *A biorthogonal system* $\{x_\lambda, f_\lambda\}$ *is a generalized basis for* X *if* $f_\lambda(x)=0$, $\lambda\in\Lambda$ *implies* $x=0$ *for all* x *in* X. $\{x_\lambda, f_\lambda\}$ *is total if the finite linear combinations of* $\{x_\lambda\}$ *are dense in* X *(i.e. if* $\{x_\lambda\}$ *is total in* X*). If* Λ *is (not) countable, such a basis is called an (extended) Markushevich basis for* X.

We note that by the condition $f_\lambda(x)=0$, $\lambda\in\Lambda\Rightarrow x=0$, $\{f_\lambda\}$ is total over X, as well as total in X^* in the weak* topology for X^* (BOURBAKI [1], III—V, p. 51).

Theorem 3. *A generalized basis for* X *is a maximal biorthogonal system with respect to* X.

Proof. If there is an x in X and an f in X^* for which $f(x)=1$ and $f_\lambda(x)=0$ on Λ we have indeed $x=0$. Thus $f(x)=0$ and this contradiction verifies the theorem.

We observe that a maximal biorthogonal system may fail to yield a generalized basis. This is shown in the last example of paragraph 3. As in the case of a basis we often refer to the set $\{x_\lambda\}$ as a generalized basis for X and $\{f_\lambda\}$ will be called the *family of coefficient functionals* corresponding to $\{x_\lambda\}$.

Let now $D(\Lambda)$ be the linear space of all scalar valued functions on Λ in which a function z is said to be zero if and only if $z(\lambda)=0$ for every λ in Λ. Let the *coefficient mapping* $F:X\to D(\Lambda)$ be the linear transformation defined by $F(x)=\{f_\lambda(x)\}$. Then the following theorem is evident from the definition of a generalized basis.

Theorem 4. *A biorthogonal system* $\{x_\lambda, f_\lambda\}$ *is a generalized basis for* X *if and only if* F *is one-to-one.*

An important consequence of this theorem is the uniqueness of the basis elements of a generalized basis.

Theorem 5. *A given family* $\{f_\lambda \mid \lambda\in\Lambda\}$ *in* X^* *can be the family of coefficient functionals for at most one generalized basis* $\{x_\lambda\}$ *for* X.

Proof. If there is another generalized basis, $\{x'_\lambda, f_\lambda\}$ for X, then $f_\lambda(x'_\mu)=\delta_{\lambda\mu}=f_\lambda(x_\mu)$, $\lambda, \mu\in\Lambda$. Hence $x'_\mu=x_\mu$ for all $\mu\in\Lambda$, since $f_\lambda(x'_\mu-x_\mu)=f_\lambda(x'_\mu)-f_\lambda(x_\mu)=0$, $\lambda\in\Lambda$.

Theorem 6. (ARSOVE) *A generalized basis* $\{x_\lambda\}$ *for a locally convex linear topological space* X *has a unique family* $\{f_\lambda\}$ *of coefficient functionals if and only if it is total.*

Proof. Given a total generalized basis $\{x_\lambda, f_\lambda\}$ for X which has another family $\{f'_\lambda\}$ of coefficient functionals, distinct of $\{f_\lambda\}$, then $f'_\lambda(x_\mu)=f_\lambda(x_\mu)=\delta_{\lambda\mu}$, $\lambda, \mu\in\Lambda$. Thus $(f'_\lambda-f_\lambda)(x)=0$ for every fixed λ in Λ

and every x which is a finite linear combination of x_μ's. From the totalness of $\{x_\lambda\}$ and the continuity of $f'_\lambda - f_\lambda$ we deduce that $(f'_\lambda - f_\lambda)(x) = 0$ for each x in X, hence that $\{f_\lambda\}$ is unique. On the other hand, suppose that $\{f_\lambda\}$ is unique, but that $\{x_\lambda\}$ is not total with respect to X. Then, by Theorem I.3.6 there is a non-zero f in X^* such that $f(x_\lambda) = 0$ on Λ. Let v be a fixed element of Λ. $(f_\lambda + \delta_{\lambda v} f)(x_\mu) = \delta_{\lambda \mu}$ shows that $\{x_\lambda, f_\lambda + \delta_{\lambda v} f\}$ is a biorthogonal system with respect to X. Assuming $(f_\lambda + \delta_{\lambda v} f)(x) = 0$ for some x in X we infer that $f_\lambda(x) = -f(x) f_\lambda(x_v)$ $= f_\lambda(-f(x) x_v)$, or in other words, $F(x) = F(-f(x) x_v)$. But since F is one-to-one one has $x = -f(x) x_v$. Hence $f(x) = f(-f(x) x_v) = -f(x) f(x_v)$ $= 0$ and $f_\lambda(x) = 0$, $\lambda \in \Lambda$, which implies that $x = 0$. Thus beside $\{x_\lambda, f_\lambda\}$ there is another generalized basis for X, $\{x_\lambda, f_\lambda + \delta_{\lambda v} f\}$, distinct from $\{x_\lambda, f_\lambda\}$ and this contradiction leads to the conclusion that the family $\{x_\lambda\}$ is total in X.

Theorem 7. (MARKUSHEVICH) *There exists a Markushevich basis for every separable Banach space X.*

Proof. By hypothesis there is a total sequence $\{x_i\}$ in X in which every finite subset of elements is linearly independent. According to the lemma of BANACH-MAZUR (IV.3.9) there exists an isometric isomorphism T of X into the Banach space $C[0,1]$. Evidently, $\overline{sp}\{Tx_i\}$ $= T(X)$. By the orthogonalization process used in the proof of Theorem VI.1.4 to construct an orthonormal basis for a separable Hilbert space H from a total sequence for H, one obtains from $\{Tx_i\}$ likewise an orthonormal set, $\{y_i\}$, in $C[0,1]$ if the inner product in $C[0,1]$ is defined by $(y, y') = \int_0^1 y(t)\overline{y'(t)}dt$, $y, y' \in C[0,1]$. Of course, in this case, the sequence $\{y_i\}$ derived from $\{Tx_i\}$ is also total, but must not necessarily be a basis for $T(X)$. Now, let $z_i = T^{-1} y_i$ and let the sequence $\{z_i^*\}$ in X^* be defined by $z_i^*(x) = (Tx, y_i)$, $x \in X$, observing that $\|z_i^*\|$ $\leq \|T\| \|y_i\|$. Because $z_i^*(z_j) = (TT^{-1} y_j, y_i) = (y_j, y_i) = \delta_{ij}$, $\{z_i, z_i^*\}$ is a biorthogonal system for X and, obviously, one has $\overline{sp}\{z_i\} = X$. Assuming now $z_i^*(x) = 0$ for all i and a given x in X, one has $(Tx, y_i) = z_i^*(x) = 0$ for all i. Since for every $\varepsilon > 0$ there is a set $\alpha_1, ..., \alpha_n$ in Φ such that $\left\| Tx - \sum_{i \leq n} \alpha_i y_i \right\| < \varepsilon$, one has

$$(Tx, Tx) = (Tx, Tx) - \sum_{i \leq n} |(Tx, y_i)|^2$$

$$\leq (Tx, Tx) - \sum_{i \leq n} |(Tx, y_i)|^2 + \sum_{i \leq n} |\alpha_i - (Tx, y_i)|^2$$

$$= (Tx - \sum_{i \leq n} \alpha_i y_i, \quad Tx - \sum_{i \leq n} \alpha_i y_i)$$

$$\leqslant \left\| Tx - \sum_{i \leqslant n} \alpha_i y_i \right\|^2$$

$$< \varepsilon^2.$$

The arbitrariness of ε implies that $Tx(t)=0$ almost everywhere on $[0,1]$ and, because this function is continuous, that $Tx=0$. From the fact that T is one-to-one one infers that $x=0$. Thus $\{z_i, z_i^*\}$ is a Markushevich basis for X and the theorem now follows.

Definition 8. *A sequence* $\{x_i\}$ *in a linear topological space* X *is a basis for* X *if for each* x *in* X *there is a unique sequence* $\{\alpha_i\}$ *in* Φ *such that* $x = \lim_n \sum_{i \leqslant n} \alpha_i x_i$ *in the topology of* X.

Evidently, each expansion coefficient α_i, by $f_i(x)=\alpha_i$, defines a linear functional f_i on X. However, as is shown later in section 5, the coefficient functionals f_i must not necessarily be continuous.

Definition 9. *A basis for a linear topological space* X *with continuous coefficient functionals* f_i, *defined by* $f_i(x)=\alpha_i$, $x \in X$, $i=1,2,\dots$ *is called a Schauder basis for* X.

Theorem 10. *Every Schauder basis for* X *is a Markushevich basis for* X. *Conversely, a Markushevich basis* $\{x_i, f_i\}$ *for* X *is a Schauder basis for* X *if and only if* $x = \lim_n \sum_{i \leqslant n} f_i(x) x_i$ *for every* x *in* X.

Proof. Let $\{x_i, f_i\}$ be a Schauder basis for X. The first assertion follows from the obvious fact that the sequence $\{x_i\}$ is total in X and from the implication $f_i(x)=0$, $i=1,2,\dots \Rightarrow x = \lim_n \sum_{i \leqslant n} f_i(x) x_i = 0$. On the other hand, let $\{x_i, f_i\}$ be a Markushevich basis such that $x = \lim_n \sum_{i \leqslant n} f_i(x) x_i$ for each $x \in X$. In this case, for each x in X, the sequence $\{f_i(x)\}$ of expansion coefficients in Φ is unique. This follows from the biorthogonality property $f_i(x_j)=\delta_{ij}$: Let $\{\beta_i\}$ be another set of expansion coefficients for x. Then one has $0 = f_i \left[\lim_n \sum_{j \leqslant n} (f_j(x) - \beta_j) x_j \right] = f_i(x) - \beta_i$ for all i.

This completes the proof of the theorem.

The following theorem gives, in the form of a counter-example, the solution of the basis problem for locally convex spaces.

Theorem 11. *If the weak* topology is assigned to* l_∞^*, *then* l_∞^* *is a locally convex separable linear topological space which has no basis.*

Proof. By (I.3.22), $J(l_1)$ is weak* dense in $l_1^{**}(=l_\infty^*)$. Since $J(l_1)$ is separable (I.4.b), and hence separable for the weak* topology of l_∞^*, it follows that l_∞^*, endowed with its (locally convex) weak* topology, is separable.

On the other hand, if one assumes that l_∞^* has a weak* basis $\{x_i^*\}$, then to every x^* in l_∞^* there is a unique sequence $\{\alpha_i\}$ in Φ such that $x^* = \lim_n \sum_{i \le n} \alpha_i x_i^*$ in the weak* topology of l_∞^*. But by (I.4.c) it is apparent that the series converges weakly to x^*. Thus $\{x_i^*\}$ is a weak basis for l_∞^* which, since l_∞^* is a Banach space, turns out to be a basis for l_∞^* (III.2.5). Thus l_∞^*, and so also l_∞ (I.3.11) would be separable in its strong topologies. This contradiction (I.4.c) completes the argument.

2. Dual Generalized Bases

Let X, X^*, F and Λ have the same meaning as defined at the beginning of the preceding section. There are then two types of biorthogonal systems $\{x_\lambda, f_\lambda\}$:
(i) $\{x_\lambda, f_\lambda\}$ is such that F is one-to-one.
(ii) $\{x_\lambda, f_\lambda\}$ is such that $\{x_\lambda\}$ is total in X (i.e. $\overline{\mathrm{sp}}\{x_\lambda\} = X$).
It follows that the set of all biorthogonal systems of type (i) are the generalized bases and that the set of all biorthogonal systems satisfying (i) and (ii) are the (extended) Markushevich bases. As we shall see, the biorthogonal systems of type (ii) are in some sense dual to those of type (i).

Definition 1. *A biorthogonal system satisfying* (ii) *is called a dual generalized basis for X.*

Theorem 2. (KLEE) *If X is separable, locally convex and* HAUSDORFF *then there exists a dual generalized basis for X.*

Proof. Since X is separable there is a sequence $\{y_i\}$ with $\overline{\mathrm{sp}}\{y_i\} = X$, and without loss of generality $\{y_i\}$ may be assumed such that every finite subset of it is linearly independent. Let N_n be the linear subspace of X spanned by $\{y_1, \ldots, y_n\}$. First of all, let $x_1 = y_1$ and $x_2 = y_2$. Then there are linear functionals f_1 and f_2 in N_2^* with the properties $f_1(x_1) = 1$, $f_1(x_2) = 0$ and $f_2(x_2) = 1$, $f_2(x_1) = 0$ (I.1.2 and I.3.6). By (I.3.5) both f_1 and f_2 have continuous linear extensions x_1^* and x_2^*, respectively, onto all of X. Thus $x_1^*, x_2^* \in X^*$ and $x_i^*(x_j) = \delta_{ij}$, $i, j = 1, 2$. We now proceed recursively. Let $\{x_1, \ldots, x_n, x_1^*, \ldots, x_n^*\}$ be such that $\mathrm{sp}\{x_1, \ldots, x_n\} = N_n$ and that $x_i^*(x_j) = \delta_{ij}$, $i, j \le n$. We observe that by the biorthogonality relations, the set $\{x_1, \ldots, x_n\}$ is linearly independent. Then we define $x_{n+1} = y_{n+1} - \sum_{i \le n} x_i^*(y_{n+1}) x_i$, and it is clear that the set $\{x_1, \ldots, x_{n+1}\}$ is linearly independent and that $\mathrm{sp}\{x_1, \ldots, x_{n+1}\} = N_{n+1}$. Finally, as above, we find $x_{n+1}^* \in X^*$ such that $x_{n+1}^*(x) = 0$, $x \in N_n$, $x_{n+1}^*(x_{n+1}) = 1$. Therefore, by a simple computation, $x_i^*(x_j) = \delta_{ij}$, $i, j \le n+1$. This shows that by induction one can obtain a biorthogonal system $\{x_i, x_i^*\}$ for X with $\overline{\mathrm{sp}}\{x_i\} = X$, which is a dual generalized basis for X.

The following theorem appears to be dual to Theorem 1.6.

Theorem 3. *A dual generalized basis* $\{x_\lambda, f_\lambda\}$ *for a locally convex linear topological space X has a unique family of basis elements $\{x_\lambda\}$ if and only if the corresponding coefficient mapping is one-to-one (i.e. if and only if $\{x_\lambda, f_\lambda\}$ is also a generalized basis).*

Proof. Let $\{x_\lambda'\}$ be another family of basis elements which is distinct of $\{x_\lambda\}$, but has the same family of coefficient functionals. Then $F(x_\mu' - x_\mu) = \{f_\lambda(x_\mu' - x_\mu)\} = \{\delta_{\lambda\mu} - \delta_{\lambda\mu}\} = 0$ for all $\mu \in \Lambda$. The uniqueness now follows assuming that F is one-to-one.

If, on the other hand, we make the hypothesis that F is not one-to-one we can show that the family $\{x_\lambda\}$ is not unique. By this hypothesis, there is a non-zero x in X such that $Fx = 0$, hence that $f_\lambda(x) = 0$ on Λ. Let then v be a fixed element of Λ. The equations $f_\lambda(x_\mu + \delta_{\mu\nu} x) = \delta_{\lambda\mu}$, $\lambda, \mu \in \Lambda$, show that $\{x_\lambda + \delta_{\lambda\nu} x, f_\lambda\}$ is a biorthogonal system with respect to X. Moreover, since $x \in X$ and since $\overline{\mathrm{sp}}\{x_\lambda\} = X$ one has, as we see later, $\overline{\mathrm{sp}}\{x_\lambda + \delta_{\lambda\nu} x\} = X$. This shows that there is another dual generalized basis for X, $\{x_\lambda + \delta_{\lambda\nu} x, f_\lambda\}$, which is distinct from $\{x_\lambda, f_\lambda\}$, hence that $\{x_\lambda\}$ is not unique. Thus, assuming $\{x_\lambda\}$ to be unique it follows that F is one-to-one and this is the desired result.

It remains to show that $\overline{\mathrm{sp}}\{x_\lambda + \delta_{\lambda\nu} x\} = X$. Let N be a given neighborhood of 0 in X. Since X is locally convex and f_ν is continuous, N contains a convex circled (Theorem I.1.4) neighborhood N' of 0 such that $f_\nu(N') \subset \{\alpha \mid |\alpha| < 1, \alpha \in \Phi\}$. Let now y be an arbitrary fixed element in X. Since $\overline{\mathrm{sp}}\{x_\lambda\} = X$ there is a linear combination $\sum_{\lambda \in \Lambda_y} \beta_\lambda x_\lambda$, where Λ_y is a finite subset of Λ containing v, such that $2\left(y - \sum_{\lambda \in \Lambda_y} \beta_\lambda x_\lambda\right) \in N'$.

On the same reasons there is a finite linear combination $\sum_{\lambda \in \Lambda_x} \alpha_\lambda x_\lambda$ such that $2(1 + |\beta_\nu|)\left(x - \sum_{\lambda \in \Lambda_x} \alpha_\lambda x_\lambda\right) \in N'$. From the assumptions it follows that $|\alpha_\nu| = \left|f_\nu\left(x - \sum_{\lambda \in \Lambda_x} \alpha_\lambda x_\lambda\right)\right| < (1 + |\beta_\nu|)^{-1}$ if $v \in \Lambda_x$. We have with $\Lambda_{xy} = \Lambda_x \cup \Lambda_y$ and the convention that $\alpha_\lambda(\beta_\lambda) = 0$ for $\lambda \notin \Lambda_x$ ($\notin \Lambda_y$, respectively),

$$\sum_{\lambda \in \Lambda_{xy}} \gamma_\lambda \left(x_\lambda + \delta_{\lambda\nu} \sum_{\mu \in \Lambda_{xy}} \alpha_\mu x_\mu\right) = \sum_{v \neq \lambda \in \Lambda_{xy}} (\beta_\lambda - \gamma_\nu \alpha_\lambda) x_\lambda + \gamma_\nu \left(x_\nu + \sum_{\lambda \in \Lambda_{xy}} \alpha_\lambda x_\lambda\right)$$

$$= \sum_{\lambda \in \Lambda_{xy}} \beta_\lambda x_\lambda,$$

9*

where $\gamma_v = (1+\alpha_v)^{-1}\beta_v$ and $\gamma_\lambda = \beta_\lambda - \gamma_v \alpha_\lambda$, $\lambda \neq v$, turn out to be finite scalars. But then,

$$y - \sum_{\lambda \in \Lambda_{xy}} \gamma_\lambda(x_\lambda + \delta_{\lambda v}x) = y - \sum_{\lambda \in \Lambda_{xy}} \gamma_\lambda \left(x_\lambda + \delta_{\lambda v} \sum_{\mu \in \Lambda_{xy}} \alpha_\mu x_\mu\right)$$
$$- \gamma_v \left(x - \sum_{\mu \in \Lambda_{xy}} \alpha_\mu x_\mu\right)$$
$$= y - \sum_{\lambda \in \Lambda_{xy}} \beta_\lambda x_\lambda - \gamma_v \left(x - \sum_{\lambda \in \Lambda_{xy}} \alpha_\lambda x_\lambda\right).$$

Since $|\gamma_v| \leqslant |\beta_v|(1 - |\alpha_v|)^{-1} < 1 + |\beta_v|$ one gets $2\gamma_v \left(x - \sum_{\lambda \in \Lambda_{xy}} \alpha_\lambda x_\lambda\right) \in N'$.
As a consequence of the convexity of N', it finally follows that $y - \sum_{\lambda \in \Lambda_{xy}} \gamma_\lambda(x_\lambda + \delta_{\lambda v}x) \in N'$ and the statement $\overline{\mathrm{sp}}\{x_\lambda + \delta_{\lambda v}x\} = X$ is verified.

3. Examples

Example 1. *Let* $B(\Lambda)$ *be the Banach space of all bounded functions* $x: \Lambda \to \Phi$, *the norm being* $\|x\| = \sup\{|x(\lambda)| \mid \lambda \in \Lambda\}$. *For every* λ *in* Λ *we define* $x_\lambda = \chi_\lambda$, *where* $\chi_\lambda(\mu) = \delta_{\lambda\mu}$ *is the characteristic function of* λ. *The functionals* f_λ, *defined by* $f_\lambda(x) = x(\lambda)$ *are obviously bounded, hence continuous, and the set* $\{x_\lambda, f_\lambda\}$ *is biorthogonal.* $\{x_\lambda, f_\lambda\}$ *is a generalized basis for* $B(\Lambda)$, *since in addition to the properties shown above,* $f_\lambda(x) = 0$ *on* Λ *implies* $x(\lambda) = 0$ *on* Λ, *hence* $x = 0$.

Example 2. *In the preceding example we take* Λ *to be the set of positive integers. Then* $B(\Lambda) = l_\infty$, *and though* l_∞ *is non-separable, it has a countable generalized basis.*

Example 3. *Let* Y *be the subspace of* l_∞ *of all elements* x *such that* $\lim_n (1/n) \sum_{\lambda \leqslant n} x(\lambda)$ *exists. It is clear that* Y *is closed in* l_∞ *and that* $\{x_\lambda, f_\lambda\}$ *of the foregoing examples provides a generalized basis for* Y. *In* Y, *the functional* f *on* Y, *used in the proof of Theorem 1.6 may be given explicitly: If* $f: Y \to \Phi$ *is defined by* $f(x) = \lim_n (1/n) \sum_{\lambda \leqslant n} x(\lambda)$, *then* f *is evidently nonzero, but vanishes at each* x_λ. *Thus* $\{x_\lambda, f_\lambda\}$ *is not a dual generalized basis for* Y.

Example 4. *Let* $A(D)$ *be the Banach space described in I.4.f. If* z_0 *is any real number such that* $0 < z_0 < 1$, *every* x *in* $A(D)$ *has the Taylor series expansion*

$$x(z) = \sum_{n=0}^{\infty} \frac{x^{(n)}(z_0)}{n!}(z - z_0)^n \qquad (1)$$

in $\{z \mid |z-z_0| < 1-z_0\} \subset D$. Let now $\{x_n\}$ be a sequence of functions in $A(D)$, given by

$$x_n(z) = (z-z_0)^n, \qquad n = 0, 1, 2, \ldots \qquad (2)$$

and let $\{f_n\}$ be a sequence of linear functionals on $A(D)$, defined by

$$f_n(x) = \frac{x^{(n)}(z_0)}{n!}, \qquad n = 0, 1, 2, \ldots. \qquad (3)$$

That the functionals f_n are continuous becomes clear by the Cauchy estimates $|x^{(n)}(z_0)| = n! \sup\{|x(z)| \mid |z-z_0| \leqslant 1-z_0\}/(1-z_0)^n$ from which follows $\|f_n\| \leqslant (1-z_0)^{-n}$. Since $f_n(x_m) = \delta_{nm}$ and since $f_n(x) = 0$, $n = 0, \ldots$, implies $x(z) = 0$ in $\{z \mid |z-z_0| < 1-z_0\}$ and, by analytic continuation, in the whole of D, the system $\{x_n, f_n\}$ is a *generalized basis* for $A(D)$. However, $\{x_n, f_n\}$ is *not a Schauder basis* for $A(D)$, because $\sum\limits_{n=0}^{\infty} (z-z_0)^n$ is the Taylor series of the function $(1+z_0-z)^{-1}$ in the open disc $\{z \mid |z-z_0| < 1\}$ and because this series is unique (Theorem I.4.13), but not convergent at the point $z = z_0 - 1$ in D.

Example 5. Let $\{D_k\}$ be a sequence of closed discs in $\mathbb{C}$, given by $D_k = \{z \mid |z| \leqslant 1-2^{-k}\}$, $k = 1, 2, \ldots$ and let each $C(D_k)$ be the Banach space of continuous complex functions on D_k. In the linear space $AF(D)$ over the field $\mathbb{C}$, of all holomorphic functions on $D = \{z \mid |z| < 1\}$, the function $\| \ \|: AF(D) \to \mathbb{R}$, given by $\|x\| = \sum\limits_{k=1}^{\infty} 2^{-k}\|x\|_k/(1+\|x\|_k)$, where $\|x\|_k$ is the norm in $C(D_k)$, describes a quasi-norm on $AF(D)$. We show that $AF(D)$ is complete in this metric and hence an F-space. For this purpose, let $\{x_n\}$ be a Cauchy sequence in $AF(D)$. Since $\|x\|_k/(1+\|x\|_k) \leqslant 2^k\|x\|$ we have $\|x\|_k \leqslant 2^{k+1}\|x\|$ for $\|x\| \leqslant 2^{-k-1}$. Since each of the spaces $C(D_k)$ is complete, $x_n(z)$ converges pointwise and uniformly on each D_k. Therefore, the limit is a function $x(z)$ which is holomorphic on D, such that $\lim\limits_{n}\|x-x_n\| = 0$, and the assertion is verified.

In the same way as in the preceding example, it can be shown that the system $\{x_n, f_n\}$ given by (2) and (3) is a *generalized basis*, but not a *Schauder basis* for $AF(D)$. The only change in the proof is for the continuity of f_n. We have $|f_n(x)| = |x^{(n)}(z_0)|/n! \leqslant \|x\|_k/(1-2^{-k}-z_0)^n$ $\leqslant 2^{k+1}\|x\|/(1-2^{-k}-z_0)^n$, if k is chosen such that $z_0 < 1-2^{-k}$ and if $\|x\|$ is taken smaller than 2^{-k-1}. As a result of WALSH [1, p. 26] the polynomials in z are dense in each $C(D_k)$. Hence $\{x_n\}$ iI total in $AF(D)$ which means that $\{x_n, f_n\}$ is a *dual generalized basis* for $AF(D)$, and, by the above, a *Markushevich basis* for $AF(D)$.

On the other hand, we take a sequence $\{x_n\}$ in $AF(D)$, given by
$$x_n(z) = \sum_{i=0}^{n} z^i, \quad n=0, 1, 2, \dots . \text{ Furthermore, the equations}$$

$$f_n(x) = \frac{x^{(n)}(0)}{n!} - \frac{x^{(n+1)}(0)}{(n+1)!}, \quad n=0, 1, \dots,$$

define a sequence of linear functionals on $AF(D)$. In quite the same way as above it can be shown that each f_n is continuous, hence an element of $AF^*(D)$. Since $f_n(x_m)=0$ for $m<n$ and for $m \geqslant n$,

$$f_n(x_m) = \left[\frac{1}{n!} \sum_{i=n}^{m} \frac{i!}{(i-n)!} z^{i-n} - \frac{1-\delta_{nm}}{(n+1)!} \sum_{i=n+1}^{m} \frac{i!}{(i-n-1)!} z^{i-n-1} \right]_{z=0}$$
$$= 1-(1-\delta_{nm}) = \delta_{nm},$$

the system $\{x_n, f_n\}$ is *biorthogonal*. To verify that $\{x_n, f_n\}$ is *maximal*, we assume the contrary. This would imply the existence of a non-zero element f of $AF^*(D)$ vanishing at each x_n. In view of the fact that each function in the space $AF(D)$ may be approximated pointwise and uniformly on each D_k by polynomials in z (WALSH [1], Theorem I.17), the sequence $\{z^n\}$, and hence also the sequence $\{x_n\}$, is total in $AF(D)$. Hence for every x in $AF(D)$ there exists a sequence $\{y_n\}$ in $AF(D)$ such that each y_n is a finite linear combination of elements in $\{x_n\}$, and such that $\lim_n y_n = x$. Thus $f(x)=\lim_n f(x-y_n)+\lim_n f(y_n)=\lim_n f(x-y_n)$ $=0$, as a consequence of the continuity of f. Since x was arbitrary we have $f=0$, which is the desired contradiction.

However, $\{x_n, f_n\}$ is *not* a *generalized basis* for $AF(D)$, since it is quite easy to find a non-zero x in $AF(D)$ for which $f_n(x)=0$ is true for all $n=0, 1, 2, \dots$: Let $x(z)=(1-z)^{-1}$. Then $f_n(x)=[(1-z)^{-n-1} -(1-z)^{-n-2}]_{z=0}=0$, $n=0, 1, \dots$, but $\|x\| \geqslant \frac{1}{2}\|x\|_1/(1+\|x\|_1)=\frac{1}{2}\cdot 2/(1+2)$ $=1/3>0$. But from the preceding considerations it is clear again that $\{x_n, f_n\}$ is a *dual generalized basis*.

4. Similar Bases

In this section we consider a special relationship between generalized bases for complete linear metric spaces with translation-invariant metric, the similarity of such bases.

Definition 1. *Let X and Y be complete linear metric spaces with translation-invariant metric, let Λ be an arbitrary index set and let $\{x_\lambda\}$, $\{y_\lambda\}$ be generalized bases for X and Y respectively, which have the same* *index set Λ. $\{x_\lambda\}$ and $\{y_\lambda\}$ are similar, if there exist families $\{f_\lambda\}$ and $\{g_\lambda\}$ of coefficient functionals for $\{x_\lambda\}$ and $\{y_\lambda\}$ respectively such that $F(X)=G(Y)$, where F and G are the coefficient mappings determined by each family of coefficient functionals.*

The definition naturally implies that both X and Y are over the same field Φ.

Theorem 2. (ARSOVE-EDWARDS) $\{x_\lambda\}$ and $\{y_\lambda\}$ are similar if and only if there exists a topological isomorphism T of X onto Y such that $y_\lambda=T\,x_\lambda,\ \lambda\in\Lambda$.

Proof. To show the sufficiency let T be a topological isomorphism of X onto Y such that $y_\lambda=T\,x_\lambda$ on Λ. Given the family of coefficient functionals $\{f_\lambda\}$ for $\{x_\lambda\}$ we choose $\{g_\lambda\}$ for $\{y_\lambda\}$ such that $g_\lambda(y)=f_\lambda(T^{-1}y)$, $y\in Y$. Since $g_\lambda(y_\mu)=f_\lambda(T^{-1}T\,x_\mu)=f_\lambda(x_\mu)=\delta_{\lambda\mu}$ and since $g_\lambda(y)=0$ on Λ implies $T^{-1}y=0$ and thus $y=0$, $\{g_\lambda\}$ is a family of coefficient functionals for $\{y_\lambda\}$ and from $f_\lambda(x)=f_\lambda(T^{-1}T\,x)=g_\lambda(T\,x)$ we have the result that $F(X)=G(Y)$.

For the necessity condition we put $Z=F(X)=G(Y)$. Observing that F and G are ono-to-one one has two possibilities of metrizing Z. Let ρ_X and ρ_Y be the metrics defined in X and Y respectively. Then one can define as a metric on Z either ρ'_X by $\rho'_X(z,0)=\rho_X(F^{-1}z,0)$ or ρ'_Y by $\rho'_Y(z,0)$ $=\rho_Y(G^{-1}z,0)$, $z\in Z$, and in both cases Z becomes a complete linear metric space (with translation-invariant metric). We can show that ρ'_X and ρ'_Y define equal topologies for Z.

Taking $\rho=\rho'_X+\rho'_Y$ it is evident that ρ, again is a translation-invariant metric on Z. If $\{z_n\}$ is a Cauchy sequence in this metric, then $\{z_n\}$ is also a Cauchy sequence in both metrics ρ'_X and ρ'_Y on Z, and there exist limits z_X and z_Y of $\{z_n\}$ in the ρ'_X and ρ'_Y topologies respectively. Since each functional f_λ is continuous we infer that $f_\lambda(F^{-1}z_X)$
$$=f_\lambda(F^{-1}(\rho'_X-)\lim_n z_n)=f_\lambda((\rho_X-)\lim_n F^{-1}z_n)=\lim_n f_\lambda(F^{-1}z_n)=\lim_n z_n(\lambda)$$
$$=\lim_n g_\lambda(G^{-1}z_n)=g_\lambda((\rho_Y-)\lim_n G^{-1}z_n)=g_\lambda(G^{-1}(\rho'_Y-)\lim_n z_n)=g_\lambda(G^{-1}z_Y)$$
$=z_Y(\lambda)=f_\lambda(F^{-1}z_Y)$ for every λ in Λ. Because $\{x_\lambda,f_\lambda\}$ is a generalized basis we infer that $F^{-1}z_X=F^{-1}z_Y$, and, since F is one-to-one, that $z_X=z_Y$. Hence $\{z_n\}$ converges in the metric ρ to the point z_X. This shows that Z is complete in the metric ρ.

Because the topologies induced by ρ'_X and ρ'_Y are both weaker than that defined by ρ, it follows from (I.1.8) that ρ'_X and ρ'_Y induce the same topology for Z. Therefore, F and G are topological isomorphisms of X and Y, respectively, onto the linear topological space Z (I.2.6). Defining T by $T=G^{-1}F$, T is the required topological isomorphism of X onto Y and, finally we have $T\,x_\lambda=G^{-1}F\,x_\lambda=G^{-1}\{f_\mu(x_\lambda)\}=G^{-1}\{\delta_{\mu\lambda}\}=y_\lambda$ which concludes the proof of the theorem.

Since T, as a topological isomorphism, preserves additional properties such as totalness of basic sequences or convergence of series in X, one has the following

Corollary 3. *If a generalized basis for a linear metric space with translation-invariant metric is an (extended) Markushevich basis or a Schauder basis, then the similar generalized bases have the same properties. Moreover, bases for Banach spaces are similar if and only if they are equivalent.*

The last statement in the corollary is a consequence of Theorem IV.3.2.

Theorem 4. *Let $\{x_\lambda\}(\lambda \in \Lambda)$ be a total generalized basis for the complete metric linear space X with translation-invariant metric $(\rho(x, y) = \rho(x - y, 0) = \|x - y\|$, $x, y \in X)$. If $\{y_\lambda\}(\lambda \in \Lambda)$ is a family of points in X and α, $0 < \alpha < 1$, is such that*

$$\left\| \sum_{i=1}^{n} \alpha_i(x_{\lambda i} - y_{\lambda i}) \right\| \leqslant \alpha \left\| \sum_{i=1}^{n} \alpha_i x_{\lambda i} \right\|$$

for all finite sequences $\lambda_1, \ldots, \lambda_n$ in Λ and all finite sequences $\alpha_1, \ldots, \alpha_n$ in Φ, then

(i) $\{y_\lambda\}$ *is a total generalized basis for X (which is similar to $\{x_\lambda\}$),*
(ii) *and there is a topological isomorphism T of X with itself such that $y_\lambda = T x_\lambda$ on Λ and $(1 - \alpha)\|x\| \leqslant \|T x\|$ on X.*

Proof. Let D be the set of all finite linear combinations of elements in the set $\{x_\lambda\}$. One has $\bar{D} = X$, since $\{x_\lambda\}$ is total in X. Let $U: D \to X$ be a linear operator defined by $U x = \sum_{\lambda \in \Lambda} f_\lambda(x)(x_\lambda - y_\lambda)$, $x \in D$, which, by the inequality of the hypothesis, is uniformly continuous. Hence U has a unique uniformly continuous extension $V: X \to X$ (I.2.5). Moreover, the hypothesis implies that $\|V^n x\| \leqslant \alpha^n \|x\|$ for all $n \geqslant 0$ and all x in X.

Now, for $S: X \to X$, given by the absolutely convergent series expansion $S x = \sum_{k=0}^{\infty} V^k x$, $x \in X$, it follows $\|S x\| \leqslant (1 - \alpha)^{-1} \|x\|$, $x \in X$. On the following reason S is one-to-one. Suppose that $S x = 0$. Due to the absolute convergence of the series for $S x$ we then have for every $\varepsilon > 0$ an index n such that $\left\| \sum_{k=0}^{m} V^k x \right\| < \varepsilon/2$ for each $m \geqslant n$, and this shows that $\|x\| \leqslant \left\| \sum_{k=0}^{n+1} V^k x \right\| + \left\| V \sum_{k=0}^{n} V^k x \right\| < \varepsilon/2 + \alpha \varepsilon/2 < \varepsilon$. Therefore, S is a continuous isomorphism of X into itself. Since $S(x - V x) = \sum_{k=0}^{\infty} V^k x - \sum_{k=0}^{\infty} V^{k+1} x = x$, it is clear that $T: X \to X$, defined by $T x = x - V x$, $x \in X$, is the (continuous) inverse of S, hence a topological isomorphism of X with itself. From the definition of V, $Tx_\lambda = y_\lambda$, $\lambda \in \Lambda$, follows immediately. If $g_\lambda : X \to \Phi$ is defined by $g_\lambda(y) = f_\lambda(T^{-1}y)$, $y \in X$, it is apparent that $\{y_\lambda, g_\lambda\}$ is a generalized basis for X which is similar to $\{x_\lambda, f_\lambda\}$. Finally, because $\overline{T(D)} = T(\overline{D}) = X$, $\{y_\lambda\}$ is total in X and the theorem is proved.

As a corollary of the above theorem one obtains the famous Paley-Wiener theorem. The theorem originally was derived in the framework of the Hilbert space L_2 in 1934 (PALEY and WIENER [1], p. 100). Then it has been generalized for Banach spaces (BOAS [1], SCHÄFKE [1]) and finally settled down for complete metric linear spaces by ARSOVE [6]. Let now X be the space defined in Theorem 4.

Theorem 5. (PALEY-WIENER) *Let $\{x_i\}$ be a Schauder basis for X. If $\{y_i\}$ is a sequence in X and α is a real number in $(0,1)$ such that*

$$\left\| \sum_{i=1}^{n} \alpha_i(x_i - y_i) \right\| \leqslant \alpha \left\| \sum_{i=1}^{n} \alpha_i x_i \right\|$$

for all finite sequences $\alpha_1, \ldots, \alpha_n$ in Φ, then
(i) *$\{y_i\}$ is a Schauder basis for X and*
(ii) *there is a topological isomorphism T on X onto itself such that $y_i = Tx_i$, $i = 1, 2, \ldots$, and $(1-\alpha)\|x\| \leqslant \|Tx\|$ on X.*

Proof. In Theorem 4 we define Λ to be the set of all positive integers and we infer that $\{y_i\}$ is a Markushevich basis for X and that (ii) holds. Due to the existence of T each element y in X then has the series expansion $\sum_{i=1}^{\infty} \alpha_i y_i$, where $\{\alpha_i\}$ is the coefficient sequence of the series expansion $\sum_{i=1}^{\infty} \alpha_i x_i$ for $T^{-1}y$. The first series is unique, since $\sum_{i=1}^{\infty} \alpha_i y_i = 0$ implies $\sum_{i=1}^{\infty} \alpha_i x_i = T^{-1} \sum_{i=1}^{\infty} \alpha_i y_i = 0$, and so $\alpha_i = 0$, $i = 1, 2, \ldots$, according to the assumption that $\{x_i\}$ is a Schauder basis for X. Hence $\{y_i\}$ is also a Schauder basis for X and we are done.

5. Continuity of the Coefficient Functionals

In this section we establish some generalizations of the theorem that every basis for a Banach space is a Schauder basis (III.1.3) and of the fact that every weak basis in a Banach space is a basis (III.2.5). We observe that the concept of a basis for a linear topological space X is, in some sense, more general than that of a generalized basis for X, in that the coefficient functionals of a basis for X may fail to be continuous. It is easy to give an example of a basis which is not a Schauder basis (cf. also to Theorem III.7.4):

Example 1. *Let Y be the space of all real functions expandable as absolutely summable power series on the interval $[0,1)$ with the topology of uniform convergence on compact subsets of $[0,1)$. Let $\{x_i\}$ be the set in Y defined by $x_i(t)=t^i$, $i=0,1,\dots$. Then $\{x_i\}$ is a basis for Y which has coefficient functionals which are not continuous.*

Proof. From the uniqueness of the expansion coefficients of in $[0,1)$ absolutely summable power series (I.4.13) it follows that $\{x_i\}$ is a basis for Y. It is clear that for each polynomial p in t, the coefficient functional f_1, determined by the expansion coefficient α_1, is given by $f_1(p)$ $=\lim_{t\to 0} t^{-1}(p(t)-p(0))$. By the following procedure we can find a sequence $\{y_n\}$ in Y such that $\lim_n y_n=0$ in the topology of Y, but such that $\lim_n f_1(y_n)=1$. Since (by I.4.5) the polynomials are dense in the Banach space $C\,[0,1]$ we can choose to each n a polynomial z_n in Y such that $\sup\{|z_n(t)-(1-nt)|\mid t\in[0,1/n]\}\leqslant 1/n$ and that $\sup\{|z_n(t)|\mid t\in(1/n,1)\}$ $\leqslant 1/n$. Obviously y_n, defined on $[0,1)$ by $y_n(t)=\int_0^t z_n(t')\,dt'$ is a polynomial. Therefore, $f_1(y_n)=\lim_{t\to 0} t^{-1} y_n(t)=z_n(0)\in[1-1/n,1+1/n]$ for $n=1,2,\dots$. But because $\sup\{|y_n(t)|\mid t\in[0,1)\}\leqslant\int_0^t |z_n(t)|\,dt\leqslant 1/(2n)+1/n=3/(2n)$, the sequence $\{y_n\}$ converges to 0 in the topology of Y and the assertion is verified.

Theorem 2. (NEWNS) *In a complete metric linear space X over $\mathbb{R}$ (or $\mathbb{C}$) which has a translation-invariant metric, every basis for X is a Schauder basis.*

Proof. Since the metric ρ in X is translation-invariant we use the notation $\|x\|=\rho(x,0)$, $x\in X$. Let $\{x_i,f_i\}$ be a basis for X. Since for each x in X, $\sum_{i\leqslant n} f_i(x)x_i$ converges to x, it is clear that $\|x\|'=\sup_n\left\|\sum_{i\leqslant n} f_i(x)x_i\right\|<\infty$. Therefore $\rho'(x,y)=\|x-y\|'$ defines a new metric on X which is stronger than ρ. Later on we shall show that X is also complete in the metric ρ'. Theorem I.1.8 then ensures that ρ and ρ' define the same topology on X. Because $\|f_n(x)x_n\|=\left\|\sum_{i\leqslant n} f_i(x)x_i-\sum_{i\leqslant n-1} f_i(x)x_i\right\|\leqslant 2\|x\|'$, and on account of the fact that $f_n(x)$ is a continuous odd function of $f_n(x)x_n$ (I.1.7), each linear functional f_n is continuous in the metric ρ', and by what has preceded, also in the metric ρ. In the following we show that X is complete in the metric ρ'.

Let $\{y_k\}$ be a Cauchy sequence in X in the metric ρ'. From the continuity of each f_n in the metric ρ we then infer that $\{f_n(y_k)\}$ is a Cauchy sequence in $\Phi = \mathbb{R}(=\mathbb{C})$, hence converges with k to some α_n in Φ. By hypothesis one has for every $\varepsilon > 0$ an integer p such that for any m and $n \geqslant m$,

$$\left\| \sum_{i=m}^{n} [f_i(y_p) - f_i(y_q)] x_i \right\| < 2\varepsilon, \qquad q \geqslant p. \tag{1}$$

Taking the limit on q one obtains

$$\left\| \sum_{i=m}^{n} \alpha_i x_i \right\| \leqslant 2\varepsilon + \left\| \sum_{i=m}^{n} f_i(y_p) x_i \right\|.$$

Since there is an index m' such that $\left\| \sum_{i=m}^{n} f_i(y_p) x_i \right\| < \varepsilon$ for all $n \geqslant m \geqslant m'$, and since X is complete in the metric ρ, $\sum_{i=1}^{n} \alpha_i x_i$ is ρ-convergent in X, say to some point y in X. Now, putting $m=1$ in (1) yields in the limit as $q \to \infty$,

$$\sup_{n} \left\| \sum_{i \leqslant n} [f_i(y_p) - \alpha_i] x_i \right\| \leqslant 2\varepsilon.$$

Therefore, $\lim_{p} \|y_p - y\|' = 0$, X is complete in the metric ρ' and by this argument the theorem is verfied.

Denote by S any compact interval in $\mathbb{R}$ and let $L_p(S)$, $0 < p < 1$, be the set of all equivalence classes of measurable functions $f : S \to \Phi$ for which $\|f\| = \int_S |f(s)|^p \, ds$ is finite. It is known (DUNFORD and SCHWARTZ [1], p. 171) that the function $\| \ \| : L_p(S) \to \mathbb{R}$ defines a quasi-norm on $L_p(S)$ and that, endowed with this quasi-norm, $L_p(S)$ is an F-space. This space now provides an interesting counter-example for the basis problem in F-spaces:

Corollary 3. $L_p(S)$, $0 < p < 1$, *are examples of separable F-spaces which have no basis.*

Proof. Since there are no nontrivial continuous linear functionals on $L_p(S)$ (DAY [1]), $L_p^*(S) = \{0\}$, $L_p(S)$ has no Schauder basis and thus, by the preceding theorem, no basis.

The separability of $L_p(S)(0 < p < 1)$ is not easy available in literature (SINGER [15], p. 454). We therefore shall scetch a way to get this result. Let first $f \in L_p(S)$ be non-negative. Then f may be approximated by a sequence $\{f_n\}$ of simple measurable functions which converges from below to f almost everywhere on S (RUDIN [1], p. 15). Since $|f(s) - f_n(s)|^p$

$\leqslant |f(s)|^p$ on S, by the Lebesgue dominated convergence theorem (I.4.9), f_n converges to f in the topology of $L_p(S)$. Next, according to Lusin's theorem (RUDIN [1], p. 53) there is for any f_n and every $\varepsilon > 0$ a function $g \in C(S)$ such that $g(s) \leqslant f_n(s)$ on S, and $g(s) = f_n(s)$ except on a set of measure $< \varepsilon$ in S. Thus $\|f_n - g\| \leqslant \varepsilon \sup\{|f_n(s)|^p \,|\, s \in S\}$, which shows, based on the separability of $C(S)$ (I.4.d), that $L_p(S)$ is also separable. The generalization to real or complex f's is familiar.

Theorem 4. *Let X be a locally convex linear topological space. Then every weak (extended) Markushevich basis for X is an (extended) Markushevich basis for X.*

Proof. The conjugate space X^* of X under the weak topology is the same as that obtained under the initial topology (I.3.3). It is thus clear that the coefficient functionals are also continuous in the initial topology of X. Since the set of basis elements in X is total in X in the weak topology of X, it is total in X in the initial topology of X (I.3.4). These properties finally show that every weak (extended) Markushevich basis for X is an (extended) Markushevich basis for X.

Theorem 5. *Let X be a barrelled (topological linear) space. Then every weak Schauder basis for X is a Schauder basis for X.*

Proof. Let $\{x_i, f_i\}$ be a weak Schauder basis for X. First of all, according to Theorem 1.10 and the preceding theorem, $\{x_i, f_i\}$ is a Markushevich basis for X. Thus $\{x_i\}$ is a total set in X which permits to choose for each x in X a sequence $\{y_n\}$ in X, converging to x, and such that $y_n \in \mathrm{sp}\{x_i | i \leqslant n\}$. As we soon will show, the family $\{T_n\}$ of continuous linear transformations of X into itself, defined by $T_n x = \sum_{i \leqslant n} f_i(x) x_i$, $x \in X$, $n = 1, 2, \ldots$, is equicontinuous in the initial topology of X. Hence

$$x = x + \lim_n T_n(x - y_n)$$
$$= x + \lim_n \sum_{i \leqslant n} f_i(x - y_n) x_i$$
$$= \lim_n y_n + \lim_n \left(\sum_{i \leqslant n} f_i(x) x_i - y_n \right)$$
$$= \lim_n \sum_{i \leqslant n} f_i(x) x_i$$

and $\{x_i, f_i\}$ is a Schauder basis for X (Theorem 1.10).

It now remains to show the equicontinuity property. Since $\lim_n T_n x = x$ in the weak topology of X for each x in X, the sequence $\{T_n x\}$ is bounded in the initial topology for each x in X (I.1.3 and I.3.4). Using the fact that X is a barrel space we invoke the Barrel theorem (I.2.4) to infer that the family $\{T_n\}$ is equicontinuous in the initial topology and this finishes the proof of the theorem.

Corollary 6. *Let X be a barrelled topological vector space and let $\{x_i, f_i\}$ be a biorthogonal system for X such that $\sup\limits_{n}\left|\sum\limits_{i \leqslant n} f(x_i) f_i(x)\right| < \infty$, $x \in X$, $f \in X^*$. Then $\{x_i, f_i\}$ is a Schauder basis for $\overline{\mathrm{sp}}\{x_i\}$.*

Proof. Evidently, $\sup\limits_{n}|f(T_n x)| < \infty$, $x \in X$, $f \in X^*$. Hence for all x, $\{T_n x\}$ is weakly bounded, and, by (I.3.4), bounded in X. The rest of the proof is analogous to that of the theorem.

References for Chapter IX: ARSOVE [4, 5], ARSOVE and EDWARDS [1], DAVIS [1], DIEUDONNÉ [1], EDWARDS [2], KLEE [1], MARKUSHEVICH [1], NEWNS [1] and SINGER [12, 15].

Bibliography

ABDELHAY, J.
[1] Caractérisation de l'espace de Banach de toutes les suites de nombres réels tendant vers zéro. C. R. Acad. Sci. Paris **229**, 1111–1112 (1949).

AKUTOWICZ, E. J.
[1] Construction of a Schauder basis in some spaces of holomorphic functions in the unit disc. Colloq. Math. **15**, 287–296 (1966).

ALAOGLU, L.
[1] Weak topologies of normed linear spaces. Ann. of Math. (2) **41**, 252–267 (1940).

ALTMAN, M. S.
[1] On biorthogonal systems. Doklady Akad. Nauk SSSR (N. S.) **67**, 413–416 (1949) (Russian). Math. Rev. **11**, 114 (1950).
[2] On bases in Hilbert space. Doklady Akad. Nauk SSSR (N. S.) **69**, 483–485 (1949) (Russian). Math. Rev. **11**, 525 (1950).

ARSOVE, M. G.
[1] The Pincherle basis problem and a theorem of Boas. Math. Scand. **5**, 271–275 (1957).
[2] Proper bases and automorphisms in the space of entire functions. Proc. Amer. Math. Soc. **8**, 264–271 (1957).
[3] Proper Pincherle bases in the space of entire functions. Quart. J. Math. (Oxford) (2) **9**, 40–54 (1958).
[4] Proper bases and linear homeomorphisms in the space of analytic functions. Math. Ann. **135**, 235–243 (1958).
[5] Similar bases and isomorphisms in Fréchet spaces. Math. Annalen **135**, 283–293 (1958).
[6] The Paley-Wiener theorem in metric linear spaces. Pacific J. Math. **10**, 365–379 (1960).

ARSOVE, M. G., and R. E. EDWARDS
[1] Generalized bases in topological linear spaces. Studia Math. **19**, 95–113 (1960).

BABENKO, K. I.
[1] On conjugate functions. Doklady Akad. Nauk SSSR (N. S.) **62**, 157–160 (1948) (Russian). Math. Rev. **10**, 149 (1949).

BANACH, S.
[1] Théorie des opérations linéaires. Warsaw, 1932.

BANACH, S., and S. MAZUR
[1] Zur Theorie der linearen Dimension. Studia Math. **4**, 100–112 (1933).
[2] Sut la divergence des séries orthogonales. Studia Math. **9**, 139–155 (1940).

BARI, N. K.
[1] Biorthogonal systems and bases in Hilbert space. Moskov Gos. Univ. Uč. Zap. 148, Matematika **4**, 69–107 (1951) (Russian). Math. Rev. **14**, 289 (1953).

BARIC, L. W.
[1] Some notes on sequences which are similar or related to a Schauder basis. Duke Math. J. **35**, 1–7 (1968).

BARIC, L. W., and W. RUCKLE
[1] Matrix transformations of Schauder bases. Studia Math. **28**, 275–278 (1966/67).

BESSAGA, C.
[1] Bases in certain spaces of continuous functions. Bull. Acad. Pol. Sci. CIII, **5**, 11–14 (1957).
[2] On topological classification of complete metric spaces. Fund. Math. **56**, 251–288 (1964/65).
[3] Topological equivalence of unseparable reflexive Banach spaces. Ordinal resolution of identity and monotone bases. Bull. Acad. Polon. Sci. Sér. Math. Astronom. Phys. **15**, 397–399 (1967).

BESSAGA, C., and A. PELCZYNSKI
[1] An extension of the Krein-Milman-Rutman theorem concering bases to the case of B_0-spaces. Bull. Acad. Pol. Sci. CIII, **5**, 379–383 (1957).
[2] On bases and unconditional convergence of series in Banach spaces. Studia Math. **17**, 151–164 (1958).
[3] A generalization of results of R. C. James concerning absolute bases in Banach spaces. Studia Math. **17**, 165–174 (1958).
[4] On subspaces of a space with an absolute basis. Bull. Acad. Polon. Sci. Sér. Sci. Math. Astr. Phys. **6**, 313–315 (1958).
[5] Properties of bases in B_0-spaces. Prace Mat. **3**, 123–142 (1959) (Polish). Math. Rev. **23**, 760 (1962).
[6] Spaces of continuous functions (IV). Studia Math. **19**, 53–62 (1960).
[7] Some remarks on homeomorphisms of Banach spaces. Bull. Acad. Polon. Sci. Sér. Sci. Math. Astronom. Phys. **8**, 757–761 (1960).

BOAS, R. P.
[1] General expansion theorems. Proc. Nat. Acad. Sci. USA **26**, 139–143 (1940).

BOČKAREV, S. V.
[1] Unconditional bases. Mat. Zametki **1**, 391–398 (1967) (Russian). Math. Rev. **35**, 135 (1968).

BONDAREV, V. G.
[1] Weak reflexivity of spaces with Schauder basis. Vestnik Moskov. Univ. Ser. I Mat. Meh. **22** no. 4, 46–49 (1967) (Russian). Math. Rev. **35**, 865 (1968).

BOURBAKI, N.
[1] Espaces vectoriels topologiques, Ch. I–V, Elements de mathématique V. Paris, 1953–55.

BUCK, R. C.
[1] Expansion theorems for analytic functions, Conference on functions of a complex variable, Univ. of Michigan, 409–419 (1953).

CALKIN, J. W.
[1] Two-sided ideals and congruences in the ring of bounded operators in Hilbert space. Ann. of Math. (2) **42**, 839–873 (1941).

ČANTURIJA, Z. A.
[1] Some properties of biorthogonal systems in Banach space and their application to spectral theory. Soobšč. Akad. Nauk Gruzin. SSR 34, 271–276 (1964)
(Russian). Math. Rev. 30, 775 (1965).
[2] On the stability of bases of Banach spaces. Soobšč. Akad. Nauk Gruzin.
SSR 36, 269–272 (1964) (Russian). Math. Rev. 30, 87 (1965).
[3] Some properties of T-bases. Soobšč. Acad. Nauk Gruzin SSR 37, 271–274
(1965) (Russian). Math. Rev. 30, 956 (1965).
[4] On some properties of biorthogonal systems in Banach space. Thbilis.
Sahelmc. Univ. Šrom. Mekh.-Makh. Mech. Ser. 110, 263–280 (1965) (Geor-
gian). Math. Rev. 33, 1356 (1967).
[5] On a problem of P. L. Ul'janov on the order of growth of the powers of a
polynomial basis. Mat. Zametki 1, 415–424 (1967) (Russian). Math. Rev. 34,
1500 (1967).

CEĬTLIN, JA. M.
[1] Unconditionality of a basis and partial order. Izv. Vysš. Učebn. Zaved.
Matematika 51, 98–104 (1966) (Russian). Math. Rev. 33, 1075 (1967).
[2] Reflexivity of spaces with a basis. Sibirsk. Math. Ž. 8, 475–479 (1967)
(Russian). Math. Rev. 35, 624 (1968).

CIESIELSKI, Z.
[1] On Haar functions and on the Schauder basis of the space C [0, 1]. Bull.
Acad. Pol. Sci. 7, 227–232 (1959).
[2] Some properties of Schauder bases of the space C [0, 1]. Bull. Acad. Polon.
Sci. 7, 141–144 (1960).
[3] Properties of the orthonormal Franklin system. Studia Math. 23, 141–157
(1963).

CIVIN, P., and B. YOOD
[1] Quasi-reflexive spaces. Proc. Amer. Math. Soc. 8, 906–911 (1957).

CUTTLE, Y.
[1] On quasi-reflexive Banach spaces. Proc. Amer. Math. Soc. 12, 936–940
(1961).

DADIČ, I.
[1] Variations of finite sets of independent vectors in Hilbert spaces. Glasnik
Mat. Ser. III 1 (21) 51—55 (1966).

DAVIS, W. J.
[1] Dual generalized bases in linear topological spaces. Proc. Amer. Math.
Soc. 17, 1057—1063 (1966).
[2] M-similarity and isomorphisms in B_0-spaces. Proc. Amer. Math. Soc. 19,
332—335 (1968).
[3] Schauder decompositions in Banach spaces. Bull. Amer. Math. Soc. 74,
1083—1085 (1968).

DAVIS, W. J., and D. W. DEAN
[1] The direct sum of Banach spaces with respect to a basis. Studia Math. 28,
209—219 (1966/67).

DAVIS, W. J., D. W. DEAN, and I. SINGER
[1] Complemented subspaces and Λ systems in Banach spaces. Israel J. Math.
6, 303—309 (1968).

DAY, M. M.
[1] The spaces L^p with $0 < p < 1$, Bull. Amer. Math. Soc. 46, 816—823 (1940).

[2] Normed linear spaces. Berlin-Göttingen-Heidelberg, 1962.
[3] On the basis problem in normed spaces. Proc. Amer. Math. Soc. **13**, 655—658 (1962).

DEAN, D. W. (see also DAVIS, W. J.)
[1] Schauder decompositions in (m). Proc. Amer. Math. Soc. **18**, 619—623 (1967).

DIEUDONNÉ, J.
[1] On biorthogonal systems. Michigan Math. J. **2**, 7—20 (1953).
[2] Foundations of modern analysis. New York, 1960.

DIXMIER, J.
[1] Sur les bases orthonormales dans les espaces préhilbertiens. Acta Sci. Math. Szeged. **15**, 29—30 (1953).

DUBINSKY, E. L., and J. R. RETHERFORD
[1] Schauder bases and Köthe sequence spaces. Bull. Acad. Polon. Sci. Sér. Math. Astronom. Phys. **14**, 497—501 (1966).
[2] Schauder bases in. compatible topologies. Studia Math. **28**, 221—226 (1966/67).

DUFFIN, R. J., and J. J. EACHUS
[1] Some notes on an expansion theorem of Paley and Wiener. Bull. Amer. Math. Soc. **48**, 850—855 (1942).

DUNFORD, N.
[1] Uniformity in linear spaces. Trans. Amer. Math. Soc. **44**, 305—316 (1938).

DUNFORD, N., and J. T. SCHWARTZ
[1] Linear operators, New York, I (1958), II (1963).

DVORETZKY, A., and C. A. ROGERS
[1] Absolute and unconditional convergence in normed linear spaces. Proc. Nat. Acad. Sci. USA **36**, 192—197 (1950).

DYNIN, A. S., and B. S. MITJAGIN
[1] Criterion for nuclearity in terms of approximative dimension. Bull. Acad. Polon. Sci. Sér. sci. math., astr. et phys. **8**, 535—540 (1960).

EACHUS, J. J. (see DUFFIN, R. J.)

EBERLEIN, W. F.
[1] Weak compactness in Banach spaces I. Proc. Nat. Acad. Sci. USA **33**, 51—53 (1947).

EDWARDS, R. E. (see also ARSOVE, M. G.)
[1] Integral bases in inductive limit spaces. Pacific J. Math. **10**, 797—812 (1960).
[2] Functional analysis, theory and applications. New York, 1965.

ELLIS, H. W., and I. HALPERIN
[1] Haar functions and the basis problem for Banach spaces. J. London Math. Soc. **31**, 28—39 (1956).

ELLIS, H. W., and D. G. KUEHNER
[1] On Schauder bases for spaces of continuous functions. Canad. Math. Bull. **3**, 173—184 (1960).

FAGE, M. K.
[1] Idempotent operators and their rectification. Doklady Akad. Nauk SSSR (N. S.) **73**, 895—897 (1950) (Russian). Math. Rev. **12**, 186 (1951).
[2] The rectification of bases in Hilbert space. Doklady Akad. Nauk SSSR (N. S.) **74**, 1053—1056 (1950) (Russian). Math. Rev. **14**, 184 (1953).

FLEMING, R. J., R. D. MCWILLIAMS, and J. R. RETHERFORD
[1] On w^*-sequential convergence, type P^* bases and reflexivity. Studia Math. **25**, 325—332 (1965).

FOGUEL, S. R.
[1] Biorthogonal systems in Banach spaces. Pacific J. Math. **7**, 1065—1072 (1957).
[2] On bases in C ([0, 1]) and L^1 ([0, 1]). Rev. Roumaine Math. Pures Appl. **10**, 931—960 (1965).

FOIAS, C., and I. SINGER
[1] Some remarks on strongly linearly independent sequences and bases in Banach spaces. Rev. math. pures appl. **6**, 589—594 (1961).

FRANKLIN, P.
[1] A set of continuous orthogonal functions. Math. Ann. **100**, 522—529 (1928).

FRINK, O.
[1] Series expansions in linear vector spaces. Amer. J. Math. **63**, 87—100 (1941).

FULLERTON, R. E.
[1] Geometric structure of absolute basis systems in a linear topological space. Pacific J. Math. **12**, 137—147 (1962).

GAPOŠKIN, V. F.
[1] On unconditional bases in the spaces L_p ($p > 1$). Uspehi Mat. Nauk (N. S.) **13**, 179—184 (1958) (Russian). Math. Rev. **20**, 1094 (1959).
[2] On certain properties of unconditional bases in the spaces L_p ($p > 1$). Uspehi Mat. Nauk (N. S.) **14**, 143—148 (1959) (Russian). Math. Rev. **22**, 988 (1961).
[3] Unconditional bases in Orlicz spaces. Uspehi Mat. Nauk **22**, no. 2 (134), 113—114 (1967) (Russian). Math. Rev. **34**, 1500 (1967).
[4] The existence of unconditional bases in Orlicz spaces. Funkcional. Anal. i Priložen. **1**, 26—32 (1967) (Russian). Math. Rev. **36**, 1100 (1968).

GARLING, D. J. H.
[1] Symmetric bases of locally convex spaces. Studia Math. **30**, 163—181 (1968).

GELBAUM, B. R.
[1] Expansions in Banach spaces. Duke Math. J. **17**, 187—196 (1950).
[2] A nonabsolute basis for Hilbert space. Proc. Amer. Math. Soc. **2**, 720—721 (1951).
[3] Notes on Banach spaces and bases. An. Acad. Brasil. Ciencias **30**, 29—36 (1958).

GELBAUM, B. R., and J. GIL DE LAMADRID
[1] Bases of tensor products of Banach spaces. Pacific J. Math. **11**, 1281—1286 (1961).

GELFAND, I. M.
[1] Normierte Ringe. Mat. Sbornik N. S. **9** (51), 3—24 (1941).
[2] Remark on the work of N. K. BARI "Biorthogonal systems and bases in Hilbert space". Moskov. Gos. Univ. Učenye Zapinski **148**, 224—225 (1951) (Russian). Math. Rev. **14**, 289 (1953).

GIL DE LAMADRID, J. (see also GELBAUM, B. R.)
[1] On finite dimensional approximation of mappings in Banach spaces. Proc. Amer. Math. Soc. **13**, 163—168 (1962).

GORDON, I. A.
[1] Certain sufficient criteria for stability of complete orthonormal bases in L [0, 1] with respect to the operation of averaging. Functional Anal. Theory of

Functions, No. I, 13—21 Izdat. Kazan. Univ., Kazan 1963 (Russian). Math. Rev. **36**, 1101 (1968).

GRINBLYUM, M. M.

[1] Certains théorèmes sur la base dans un espace du type (*B*). Doklady Akad. Nauk SSSR (N. S.) **31**, 428—432 (1941) (Russian). Math. Rev. **3**, 49 (1942).

[2] Biorthogonal systems in Banach space. Doklady Akad. Nauk SSSR (N. S.) **47**, 79—82 (1945) (Russian). Math. Rev. **7**, 125 (1946).

[3] Sur la théorie des systèmes biorthogonaux. Doklady Akad. Nauk SSSR (N. S.) **55**, 287—290 (1947).

[4] On a property of a basis. Doklady Akad. Nauk SSSR (N. S.) **59**, 9—11 (1948) (Russian). Math. Rev. **10**, 307 (1949).

[5] On the representation of a space of type *B* in the form of a direct sum of subspaces. Doklady Akad. Nauk SSSR (N. S.) **70**, 749—752 (1950) (Russian). Math. Rev. **11**, 525 (1950).

GRÜNBAUM, B.

[1] Some applications of expansion constants. Pacific J. Math. **10**, 194—201 (1960).

GURARIĬ, V. I.

[1] On inclinations of spaces and conditional bases in Banach space. Doklady Akad. Nauk SSSR **145**, 504—506 (1962) (Russian). Math. Rev. **27**, 553 (1964).

[2] Bases in spaces of continuous functions. Doklady Akad. Nauk SSSR **148**, 483—495 (1963) (Russian). Math. Rev. **26**, 556 (1963).

[3] Some geometric characteristics of subspaces and bases in Banach spaces. Colloq. Math. **13**, 59—63 (1964) (Russian). Math. Rev. **31**, 467 (1966).

[4] The index of sequences in $\tilde{C}$ and the existence of infinite dimensional separable Banach spaces having no orthogonal basis. Rev. Roumaine Math. Pures Appl. **10**, 967—971 (1965) (Russian). Math. Rev. **34**, 591 (1967).

[5] Bases for sets in Banach spaces. Rev. Roumaine Math. Pures Appl. **10**, 1235—1240 (1965).

[6] Bases in spaces of continuous functions on compacta and some geometric questions. Izv. Akad. Nauk SSSR Ser. Mat. **30**, 289—306 (1966) (Russian). Math. Rev. **34**, 591 (1967).

[7] Subspaces and bases in spaces of continuous functions. Doklady Akad. Nauk SSSR **167**, 971—973 (1966) (Russian). Math. Rev. **33**, 1358 (1967).

GURARIĬ, V. I., and M. I. KADEC

[1] Minimal systems and quasi-complements in Banach space. Doklady Akad. Nauk SSSR **145**, 256—258 (1962) (Russian). Math. Rev. **26**, 1276 (1963).

GUREVIČ, L. A.

[1] On unconditional bases. Uspehi Mat. Nauk (N. S.) **8**, no. 5 (57), 153—156 (1953) (Russian). Math. Rev. **15**, 631 (1954).

[2] A basis in the space of abstract functions. Doklady Akad. Nauk SSSR **136**, 12—15 (1961) (Russian). Math. Rev. **24**, 411 (1962).

[3] Conic tests for bases of absolute convergence. Problems of Math. Phys. and Theory of Functions, II. p. 12—21. Naukova Dumka, Kiev, 1964 (Russian). Math. Rev. **33**, 1358 (1967).

HAAR, A.

[1] Zur Theorie der orthogonalen Funktionssysteme, I, Math. Ann. **69**, 331—337 (1910), II, ibid. **71**, 38—53 (1911).

HALMOS, P. R.

[1] Finite dimensional vector spaces. Princeton, 1958.

[2] A Hilbert space problem book. Princeton, 1967.

HAUSDORFF, F.
[1] Mengenlehre. New York, 1944.

HILDEBRANDT, T. H.
[1] On unconditional convergence in normed vector spaces. Bull. Amer. Math.
Soc. **46**, 959—962 (1940).

HILDING, S. H.
[1] Note on completeness theorems of Paley-Wiener type. Ann. of Math. (2)
49, 953—955 (1948).

HILLE, E., and R. S. PHILLIPS
[1] Functional analysis and semi-groups. Amer. Math. Soc. Colloquium Publ.
31 (rev. ed.) (1957).

ISTRATESCU, V.
[1] Über die Banachräume mit zählbarer Basis I, Rev. Math. Pures Appl.
(Bucarest) **7**, 481—482 (1962), II, Revue Roumaine Math. Pures Appl. **9**,
431—433 (1964).

IYER, V. G.
[1] On the space of integral functions (III). Proc. Amer. Math. Soc. **3**, 874—883
(1952).

JAMES, R. C.
[1] Orthogonality in normed linear spaces. Duke Math. J. **12**, 291—302 (1945).
[2] Orthogonality and linear functionals in normed linear spaces. Trans. Amer.
Math. Soc. **61**, 265—292 (1947).
[3] Inner products in normed linear spaces. Bull. Amer. Math. Soc. **53**, 559—566
(1947).
[4] Bases and reflexivity of Banach spaces. Ann. of Math. **52**, 518—527 (1950).
[5] A non-reflexive Banach space isometric with its second conjugate space.
Proc. Nat. Acad. Sci. USA **37**, 174—177 (1951).
[6] Projections in the space (m). Proc. Amer. Math. Soc. **6**, 899—902 (1955).
[7] Separable conjugate spaces. Pacific J. Math. **10**, 563—571 (1960).
[8] Characterizations of reflexivity. Studia Math. **23**, 205—216 (1964).
[9] Weak compactness and reflexivity. Israel J. Math. **2**, 101—119 (1964).
[10] Weakly compact sets. Trans. Amer. Math. Soc. **113**, 129—140 (1964).

JAMES, R. C., and J. R. RETHERFORD
[1] Unconditional bases and best approximation in Banach spaces. Bull. Amer.
Math. Soc. **75**, 108—112 (1969).

JONES, O. T., and J. R. RETHERFORD
[1] On similar bases in barrelled spaces. Proc. Amer. Math. Soc. **18**, 677—680
(1967).

JULIA, G.
[1] Exemples des structures des systèmes duaux de l'espace hilbertien. C. R.
Acad. Sci. (Paris) **216**, 465—468 (1943).

KACZMARZ, S., and H. STEINHAUS
[1] Theorie der Orthogonalreihen. Warsaw, 1935.

KADEC, M. I. (see also GURARIĬ, V. I.)
[1] On conditionally convergent series in the space L_p. Uspehi Mat. Nauk
(N. S.) **11**, 107—109 (1954) (Russian). Math. Rev. **15**, 802 (1954).
[2] Linear dimension of the spaces L_p and l_p. Uspehi Mat. Nauk (N. S.) **13**,
95—98 (1958) (Russian). Math. Rev. **21**, 59 (1960).

[3] Bases and their spaces of coefficients. Dopovidi Akad. Nauk Ukrain. RSR **1**, 1139—1140 (1964).
[4] Topological equivalence of all separable Banach spaces. Soviet Math. Doklady **7**, 319—322 (1966).
[5] Nonlinear operator-bases in a Banach space. Teor. Funkciĭ Funkcional. Anal. i Priložen. Vyp. **2**, 128—130 (1966) (Russian). Math. Rev. **34**, 1188 (1967).

KADEC, M. I., and A. PELCZYNSKI
[1] Bases, lacunary sequences and complemented subspaces in the spaces L_p. Studia Math. **21**, 161—176 (1962).
[2] Basic sequences, biorthogonal systems and norming sets in Banach and Fréchet spaces. Studia Math. **25**, 297—323 (1965).

KAKUTANI, S.
[1] Some characterizations of Euklidean spaces. Jap. J. Math. **16**, 93—97 (1939).

KARLIN, S.
[1] Unconditional convergence in Banach spaces. Bull. Amer. Math. Soc. **54**, 148—152 (1948).
[2] Bases in Banach spaces. Duke Math. J. **15**, 971—985 (1948).

KELLEY, J. L., and I. NAMIOKA
[1] Linear topological spaces. Princeton, 1963.

KLEE, V.
[1] On the borelian and projective types of linear subspaces. Math. Scand. **6**, 189—199 (1958).

KÖTHE, G.
[1] Probleme der linearen Algebra in topologischen Vektorräumen. Proc. internat. sympos. on linear spaces 1960. Jerusalem, Academic Press 1961.
[2] Topologische lineare Räume. Berlin-Heidelberg-New York, 1966.

KOSTYUČENKO, A., and A. SKOHOROD
[1] On a theorem of M. K. Bari. Uspehi Mat. Nauk (N. S.) **8**, no. 5 (57), 165—166 (1953) (Russian). Math. Rev. **15**, 632 (1954).

KOZLOV, V. YA.
[1] On bases in the space L [0,1]. Mat. Sbornik N. S. **26** (68), 85—102 (1950) (Russian). Math. Rev. **11**, 602 (1950).
[2] On a generalization of the concept of a basis. Doklady Akad. Nauk SSSR (N. S.) **73**, 643—646 (1950) (Russian). Math. Rev. **12**, 110 (1951).

KREIN, M., D. MILMAN, and M. RUTMAN
[1] A note on a basis in Banach space. Comm. Inst. Sci. Math. Méc. Univ. Kharkoff (Zapinski Inst. Mat. Mech.) (4) **16**, 106—110 (1940) (Russian). Math. Rev. **3**, 49 (1942).
[2] On a property of a basis in Banach space. Kark. Zap. Matem. Obsh. (4) **16**, 182 (1940) (Russian).

KUEHNER, D. G. (see ELLIS, H. W.)

LINDENSTRAUSS, J.
[1] Extension of compact operators. Mem. Amer. Math. Soc. **48** (1964).
[2] On a subspace of the space 1. Bull. Acad. Polon. Sci., Sér. Sci. Math. Astronom. Phys. **12**, 539—542 (1964).

LINDENSTRAUSS, J., and M. ZIPPIN
[1] Banach spaces with a unique unconditional basis. J. Funct. Anal. **3**, 115—125 (1969).

LJUSTERNIK, L. A., and W. I. SOBOLEW
[1] Elemente der Funktionalanalysis. Berlin, 1955.

LOZANOVSKIĬ, G. JA.
[1] On Banach lattices and bases. Functional. Anal. i Priložen. 1, no. 3, 92 (1967) (Russian). Math. Rev. 36, 633 (1968).

MACPHAIL, M. S.
[1] Absolute and unconditional convergence. Bull. Amer. Math. Soc. 53, 121—123 (1947).

MADDAUS, I.
[1] On completely continuous linear transformations. Bull. Amer. Math. Soc. 44, 279—282 (1938).

MARCINKIEWICZ, J.
[1] Quelques théorèmes sur les séries orthogonales. Ann. Soc. Polon. Math. 16, 84—96 (1937).

MARKUS, A. S.
[1] A basis of root vectors of a dissipative operator. Soviet Math. Doklady 1, 599—602 (1960).

MARKUSHEVICH, A. I.
[1] Sur les bases (au sens large) dans les espaces linéaires. Doklady Akad. Nauk SSSR (N. S.) 41, 227—229 (1943).
[2] Sur la meilleure approximation. Doklady Akad. Nauk SSSR (N. S.) 44, 262—264 (1944).
[3] On bases in the space of analytic functions. Mat. Sbornik 17 (59), 211—252 (1945) (Russian). Math. Rev. 7, 425 (1946).

MARTI, J. T.
[1] On integro-differential equations in Banach spaces. Pacific J. Math. 20, 99—108 (1967).
[2] Extended bases for Banach spaces. To appear.
[3] On bases, compactness and weak convergence in the Banach space A_p. To appear.

MAZUR, S. (see BANACH, S.)

MAZUR, S., and W. ORLICZ
[1] Sur les espaces métriques linéaires I, II. Studia Math. I, 10, 184—208 (1948), II, 13, 137—179 (1953).

MCARTHUR, C. W.
[1] On relationships among certain spaces of sequences in an arbitrary Banach space. Canad. J. Math. 8, 192—197 (1956).
[2] The weak basis theorem. Colloq. Math. 17, 71—76 (1967).

MCARTHUR, C. W., and J. R. RETHERFORD
[1] Uniform and equicontinuous Schauder bases of subspaces. Canad. J. Math. 17, 207—212 (1965).
[2] Some remarks on bases in linear topological spaces. Math. Ann. 164, 38—41 (1966).

MCKINNEY, R. L.
[1] Positive bases for linear spaces. Trans. Amer. Math. Soc. 103, 131—148 (1962).

MCWILLIAMS (see FLEMING, R. J.)

MICHAEL, E., and A. PELCZYNSKI
[1] Separable Banach spaces which admit l_n approximations. Israel Math. J. **4**, 189—198 (1966).

MILMAN, V. D. (see also KREIN, M.)
[1] Certain properties of unconditional bases. Soviet Math. Doklady **6**, 656—659 (1965).
[2] Some properties of sequences of elements of Banach spaces. First Republ. Math. Conf. of Young Researchers, Part II, Akad. Nauk Ukrain. SSR Inst. Math., Kiev, 480—489 (1965) (Russian). Math. Rev. **34**, 885 (1967).

MITJAGIN, B. S. (see also DYNIN, A. S.)
[1] Approximative dimension and bases in nuclear spaces. Uspehi Mat. Nauk **16**, 63—132 (1961) (Russian). Math. Rev. **27**, 554 (1964).

MURRAY, F. J.
[1] On complementary manifolds and projections in spaces L_p and l_p. Trans. Amer. Math. Soc. **41**, 138—152 (1937).
[2] The analysis of linear transformations. Bull. Amer. Math. Soc. **48**, 76—93 (1942).

SZ. NAGY, B.
[1] Expansion theorems of Paley-Wiener type. Duke Math. J. **14**, 975—978 (1947).

NAMIOKA, I. (see KELLEY, J. L.)

NEWNS, W. F.
[1] On the representation of analytic functions by infinite series. Phil. Trans. Royal Soc. London (A) **245**, 429—468 (1953).

NGUYEN THANH VAN
[1] Bases de Schauder dans certains espaces vectoriels topologoques. Ann. Fac. Sci. Univ. Toulouse (4) **28**, 139—147 (1965).
[2] Sur les bases de Schauder de l'espace des fonctions holomorphes dans une domaine simplement connexe. C. R. Acad. Sci. Paris Sér. A—B **264**, A 1053 bis 1055 (1967).

NGUEN VAN KHUE
[1] Test for unconditional convergence bases. Izv. Vysš. Učebn. Zaved. Matematika **69**, 68—74 (1968) (Russian). Math. Rev. **36**, 1099 (1968).

NIKOL'SKIĬ, V. N.
[1] The best approximation and a basis in a Fréchet space. Doklady Akad. Nauk SSSR (N. S.) **59**, 639—642 (1948) (Russian). Math. Rev. **10**, 128 (1949).
[2] Some questions of best approximation in a function space. Uč. Zap. Kalininsk. Pedagog. Inst. **16**, 119—160 (1954) (Russian). Math. Rev. **17**, 175 (1956).

OLUBUMMO, A.
[1] Operators of finite rank in a reflexive Banach space. Pacific J. Math. **12**, 1023—1027 (1962).

ORLICZ, W. (see also MAZUR, S.)
[1] Beiträge zur Theorie der Orthogonalreihenentwicklungen II. Studia Math. **1**, 241—255 (1929).
[2] Über unbedingte Konvergenz in Funktionenräumen I. Studia Math. **4**, 33—37 (1933).
[3] Some remarks on the absolute convergence of biorthogonal expansions in the space C. Ann. Univ. Sci. Budapest Eötvös Sect. Math. **3**—**4**, 217—222 (1960/61).

PALEY, R. E. A. C., and N. WIENER
[1] Fourier transforms in the complex domain. Amer. Math. Soc. Colloquium Publ. no. 19, New York, 1934.

PALEY, R. E. A. C., and A. ZYGMUND
[1] On some series of functions (I). Proc. Cambr. Phil. Soc. 26, 337—357 (1930).

PECK, N. T.
[1] On non locally convex spaces II. Math. Ann. 178, 209—218 (1968).

PELCZYNSKI, A. (see also BESSEGA, C., KADEC, M. I., and MICHAEL, E.)
[1] On B-spaces containing subspaces isomorphic to the space c_0. Bull. Acad. Pol. Sci. Cl. III, 5, 797—798 (1957).
[2] A connection between weakly unconditional convergence and weak completeness of Banach spaces. Bull. Acad. Polon. Sci. Sér. Math. Astr. Phys. 6, 251—253 (1958).
[3] Projections in certain Banach spaces. Studia Math. 19, 209—228 (1960).
[4] A note to the paper of I. Singer ,,Basic sequences and reflexivity of Banach spaces". Studia Math. 21, 371—374 (1962).
[5] Some problems on bases in Banach and Fréchet spaces. Israel J. Math. 2, 132—138 (1964).
[6] On simultaneous extension of continuous functions. Studia Math. 24, 285—304 (1964).
[7] A proof of the Eberlein-Smulian theorem by an application of basic sequences. Bull. Acad. Polon. Sci., Sér. Sci. Math. astr. phys. 12, 543—548 (1964).

PELCZYNSKI, A., and I. SINGER
[1] On non-equivalent bases and conditional bases in Banach spaces. Studia Math. 25, 5—25 (1964).

PELCZYNSKI, A., and W. SZLENK
[1] An example of a non-shrinking basis. Rev. Roumaine Math. Pures Appl. 10, 961—966 (1966).

PETTIS, B. J.
[1] On integration in vector spaces. Trans. Amer. Math. Soc. 44, 227—304 (1938).

PHILLIPS, R. S. (see also HILLE E.)
[1] On linear transformations. Trans. Amer. Math. Soc. 48, 516—541 (1940).

PIETSCH, A.
[1] Nukleare lokalkonvexe Räume. Berlin, 1965.
[2] F-Räume mit absoluter Basis. Studia Math. 26, 233—238 (1966).

POLLARD, H.
[1] Completeness theorems of Paley-Wiener type. Ann. of Math 45, 738—739 (1944).

POLTAVSKIĬ, L. N.
[1] Orthogonality in L_p-spaces. Vestnik Moskov Univ. Ser. I Mat. Meh. 22, no. 1, 47—50 (1967) (Russian). Math. Rev. 35, 139 (1968).

PRIGORSKIĬ, V. A.
[1] On some classes of bases in Hilbert space. Uspehi Mat. Nauk 20, no. 5 (125), 231—236 (1965) (Russian). Math. Rev. 34, 317 (1967).
[2] On quadratically stable bases of subspaces. Mat. Issled. 2, 164—168 (1967) (Russian). Math. Rev. 36, 633 (1968).

PTÁK, V.
[1] Biorthogonal systems and reflexivity of Banach spaces. Czechosl. Math. J. 9, 319—326 (1959).

RADEMACHER, H.
[1] Einige Sätze über Reihen von allgemeinen Orthogonalfunktionen. Math. Ann. **87**, 112—138 (1922).

REAY, J. R.
[1] Unique minimal representations with positive bases. Amer. Math. Monthly **73**, 253—261 (1966).

RETHERFORD, J. R. (see also DUBINSKY, E. L., JAMES, R. C., JONES, O. T., FLEMING, R. J., and MCARTHUR, C. W.)
[1] Basic sequences and the Paley-Wiener criterion. Pacific. J. Math. **14**, 1019—1027 (1964).
[2] w^*-bases and bw^*-bases in Banach spaces. Studia Math. **25**, 65—71 (1964).
[3] Bases, basic sequences and reflexivity of linear topological spaces. Math. Ann. **164**, 280—285 (1966).
[4] Shrinking bases in Banach spaces. Amer. Math. Monthly **73**, 841—846 (1966).
[5] Some remarks on Schauder bases of subspaces. Rev. Roumaine Math. Pures Appl. **11**, 787—792 (1966).
[6] On Čebyšev subspaces and unconditional bases in Banach spaces. Bull. Amer. Math. Soc. **73**, 238—241 (1967).
[7] Some characterizations of c_0 and l^1. Canad. Math. Bull. **10**, 39—52 (1967).

RICKART, C. E.
[1] General theory of Banach algebras. Princeton, 1960.

ROGERS, C. A. (see DVORETZKY, A.)

RUCKLE, W. H. (see also BARIC, L. W.)
[1] Schauder decompositions and bases. Dissertation, Florida State Univ., 1963.
[2] The infinite sum of closed subspaces of an F-space. Duke Math. J. **31**, 543—554 (1964).
[3] Infinite matrices which preserve Schauder bases. Duke Math. J. **33**, 547—550 (1966).
[4] On the construction of sequence spaces that have Schauder bases. Canad. J. Math. **18**, 1281—1293 (1966).
[5] On the characterization of sequence spaces associated with Schauder bases. Studia Math. **28**, 279—288 (1966/67).
[6] Lattices of sequence spaces. Duke Math. J. **35**, 491—503 (1968).

RUDIN, W.
[1] Real and complex analysis. New York, 1966.

RUSSO, J. P.
[1] Monotone and e-Schauder bases of subspaces. Canad. J. Math. **20**, 233—241 1968).

RUTMAN, M. (see KREIN, M.)

RUTOWITZ, D.
[1] Absolute and unconditional convergence in normed linear spaces. Proc. Cambr. Phil. Soc. **58**, 575—579 (1962).

ŠAĬDUKOV, K. M.
[1] On the Lebesgue constants of bases in the space of continuous functions. Functional Anal. Theory of Functions no. I, Izdat. Kazan Univ. Kazan, 122—133 (1963) (Russian). Math. Rev. **34**, 1190 (1967).

[2] On the order of growth of the degrees of a polynomial basis. Functional Anal. Theory of Functions no. I, Izdat. Kazan Univ. Kazan, 134—138 (1963) (Russian). Math. Rev. **34**, 1501 (1967).

[3] A criterion for a basis in the space of continuous functions. Izv. Vysš. Učebn. Zaved Matematica **52**, 178—182 (1966) (Russian). Math. Rev. **33**, 302 (1967).

SANDERS, B. L.

[1] On a generalization of the Schauder basis concept. Dissertation the Florida State University (1962).

[2] Decompositions and reflexivity in Banach spaces, Proc. Amer. Math. Soc. **16**, 204—208 (1965).

[3] On the existence of (Schauder) decompositions in Banach spaces, Proc. Amer. Math. Soc. **16**, 987—990 (1965).

SCHAEFER, H. H.

[1] Halbgeordnete lokalkonvexe Vektorräume. Math. Ann. **135**, 115—141 (1958).

[2] Topological vector spaces. New York, 1966.

SCHÄFFER, J. J.

[1] Another characterization of Hilbert spaces. Studia Math. **25**, 271—276 (1965).

SCHÄFKE, F. W.

[1] Das Kriterium von Paley und Wiener im Banachschen Raum. Math. Nachr. **3**, 59—61 (1949).

SCHAUDER, J.

[1] Zur Theorie stetiger Abbildungen in Funktionalräumen, Math. Z. **26**, 47—65 (1927).

[2] Eine Eigenschaft des Haarschen Orthogonalsystems, Math. Z. **28**, 317—320 (1928).

SCHMIDT, E.

[1] Entwicklung willkürlicher Funktionen nach Systemen vorgeschriebener. Math. Ann. **63**, 433—476 (1907).

SCHWARTZ, J. T. (see DUNFORD, N.)

SEMADENI, Z.

[1] Product Schauder bases and approximation with nodes in spaces of continuous functions. Bull. Acad. Polon. Sci. Sér. Sci. Math. Astronom. Phys. **11**, 387—391 (1963).

SINGER, I. (see also DAVIS, W. J., FOIAS, C., and PELCZYNSKI, A.)

[1] Elementary proof of a theorem of S. R. Foguel on biorthogonal systems in Banach spaces. Rev. Math. pures appl. **3**, 305—307 (1958).

[2] Sur les espaces de Banach à base absolue, canoniquement équivalents à un dual d'espace de Banach, C. R. Acad. Sci. Paris **251**, 620—621 (1960).

[3] Weak* bases in conjugate Banach spaces. Studia Math. **21**, 75—81 (1961).

[4] On Banach spaces with a symmetric basis. Rev. Math. Pures et Appl. (Bucarest) **6**, 159—166 (1961) (Russian). Math. Rev. **26**, 797 (1963).

[5] Basic sequences and reflexivity of Banach spaces. Studia Math. **21**, 351—369 (1962).

[6] Some characterizations of symmetric bases, Bull. Acad. Pol. Sci., Serie math., astr. et phys. **10**, 185—192 (1962).

[7] On Cesaro bases in Banach spaces. Rev. math. pures appl. (Bucarest) 7, 135—142 (1962).

[8] On a theorem of I. M. Gelfand. Uspehi Mat. Nauk 17, 169—176 (1962) (Russian). Math. Rev. 24, 656 (1962).

[9] On Banach spaces reflexive with respect to a linear subspace of their conjugate space II, III. Math. Ann. 145, 64—76 (1962), Rev. math. pures appl. 8, 139—150 (1963).

[10] Weak* bases in conjugate Banach spaces II. Rev. Math. Pures Appl. (Bucarest) 8, 575—584 (1963).

[11] On bases in quasi-reflexive Banach spaces. Rev. Math. pures Appl. (Bucarest) 8, 309—311 (1963).

[12] Bases in Banach spaces I, II and III. Studie si cercetari matematice 14, 533—585 (1963); 15, 157—208 (1964); 15, 675—725 (1964) (Rumanian).

[13] A proof of the Dvoretzky-Rogers theorem, Israel J. Math. 2, 249—250 (1964).

[14] Bases and quasi-reflexivity of Banach spaces. Math. Ann. 153, 199—209 (1964).

[15] On the basis problem in topological linear spaces. Rev. Roumaine Math. Pures Appl. 10, 453—457 (1965).

[16] Bases in Banach spaces. Grundl. d. math. Wiss. 154, Berlin-Heidelberg-New York, in preparation.

SIRETCHI, G.
[1] On certain spaces with a base. An. Univ. Bucaresti Ser. Stiint Natur. Mat.-Mech. 13, 141—144 (1964).

SKOHOROD, A. (see KOSTYUČENKO, A.)

SOBCZYK, A.
[1] Projection of the space (m) on its subspace (c_0). Bull. Amer. Math. Soc. 47, 938—947 (1941).

SOBOLEW, W. I. (see LJUSTERNIK, L. A.)

SOLOMYAK, M. Z.
[1] On orthogonal basis in Banach space. Vestnik Leningrad. Univ. 12, 27—36 (1957) (Russian). Math. Rev. 19, 45 (1958).

STEINHAUS, H. (see KACZMARZ, S.)

SZILENK, W. (see also PELCZYNSKI, A.)
[1] Une remarque sur l'orthogonalisation des bases de Schauder dans l'espace C. Colloq. Math. 15, 297—301 (1966).

TAYLOR, A. E.
[1] The extension of linear functionals. Duke Math. J. 5, 538—547 (1939).
[2] The weak topologies of Banach spaces. Proc. Nat. Acad. Sci. USA 25, 438—440 (1939).
[3] A geometric theorem and its application to biorthogonal systems. Bull. Amer. Math. Soc. 53, 614—616 (1947).
[4] Introduction to functional analysis. New York, 1958.

TOEPLITZ, O.
[1] Über allgemeine lineare Mittelbildungen. Prace Math. Fiz. 22, 113—119 (1911).

TROJANSKIĬ, S.
[1] The topological equivalence of the spaces $c_0(\aleph)$ and $l(\aleph)$. Bull. Acad. Polon. Sci. Sér. Sci. Math. Astronom. Phys. 15, 389—396 (1967) (Russian). Math. Rev. 36, 1099 (1968).

TSENG, Y. Y.
[1]. On generalized biorthogonal expansions in metric and unitary spaces. Proc. Nat. Acad. Sci. USA **28**, 35—43 (1942).

VAHER, F. S.
[1] On the basis in the space of continuous functions defined on a compact set. Doklady Akad. Nauk SSSR **101**, 589—592 (1955) (Russian). Math. Rev. **16**, 1031 (1955).

VAĬC, B. E.
[1] On some properties of unconditioal bases. Uspehi Mat. Nauk **17** no. 6 (108), 135—142 (1962) (Russian). Math. Rev. **26**, 797 (1963).
[2] Some stability properties of bases. Soviet Math. Doklady **5**, 1141—1144 (1964).
[3] Characteristic properties of unconditional bases and theorems of stability. Izv. Vyss. Učebn Zaved Matematika **47**, 24—36 (1965) (Russian). Math. Rev. **34**, 1191 (1967).

VANIČEK, J.
[1] Biorthogonal systems in a Banach space (Engl. summary). Acta Fac. Rev. natur. Univ. Comenian. Math. **6**, 319—325 (1961).
[2] Biorthogonal systems and limit methods. Casopis Rest. Mat. **87**, 17—21 (1962) (Czech.). Math. Rev. **24**, 533 (1962).
[3] Approximating sequences in Banach spaces. Casopis Rest, Mat. **87**, 52—62 (1962) (Czech.). Math. Rev. **24**, 533 (1962).

VANIČEK, J., and H. VANIČKOVA
[1] On the space of holomorphic functions. Casopis Rest. Mat. **86**, 433—438 (1961) (Czech.). Math. Rev. **24**, 616 (1962).

VANIČKOVA, H. (see VANIČEK, J.)

VINIKUROV, V. G.
[1] On biorthogonal systems spanning a given subspace. Doklady Akad. Nauk SSSR (N. S.) **85**, 685—687 (1952) (Russian). Math. Rev. **14**, 183 (1953).

VIZITEĬ, V. N.
[1] Stability of bases consisting of subspaces of a Banach space. Studies in Algebra and Math. Anal. Izdat "Karta Malovenjaske", Kishinev, 34—44 (1965) (Russian). Math. Rev. **34**, 884 (1967).

VOLENEC, V.
[1] Variations of orthogonal basic sets in Euclidean space. Glasnik Mat. Ser. III **1** (21) 51—55 (1966).

WALSH, J. L.
[1] Interpolation and approximation by rational functions in the complex domain. Amer. Math. Soc. Colloquium Publ. **20**, 1965.

WEILL, L. J.
[1] Stability of bases in complete barrelled spaces. Proc. Amer. Math. Soc. **18**, 1045—1050 (1967).

WIENER, N. (see PALEY, R. E. A. C.)

WILANSKY, A.
[1] The basis in Banach space. Duke Math. J. **18**, 795—798 (1951).
[2] Functional analysis. New York, 1964.

WOJTYŃSKI, W.
[1] On bases in certain countably-Hilbert spaces. Bull. Acad. Polon. Sci. Sér. Sci. Math. Astronom. Phys. **14**, 681—684 (1966).

YAMAZAKI, S.
   [1] Normed rings and unconditional bases in Banach spaces. Sci. Pap. Coll. Gen. Educ. Univ. Tokyo **14**, 1—10 (1964).
   [2] Normed rings and bases in Banach spaces. Sci. Pap. Coll. Gen. Educ. Univ. Tokyo **15**, 1—13 (1965).
   [3] Remerk to "Normed rings and bases in Banach spaces". Sci. Pap. Coll. Gen. Educ. Univ. Tokyo **16**, 25—26 (1966).

YOOD, B. (see CIVIN, P.)

YOSIDA, K.
   [1] Functional analysis. Berlin-Göttingen-Heidelberg, 1965.

ZAHARJUTA, V. P.
   [1] Continuable bases in spaces of analytic functions of one and several variables. Sibirsk. Mat. Ž. **8**, 277—292 (1967) (Russian). Math. Rev. **35**, 1092 (1968).

ZIPPIN, M. (see also LINDENSTRAUSS, A.)
   [1] On a certain basis in c_0. Israel J. Math. **4**, 199—204 (1966).
   [2] On perfectly homogeneous bases in Banach spaces. Israel J. Math. **4**, 265—272 (1966).

ZYGMUND, A. (see also PALEY, R. E. A. C.)
   [1] Trigonometrical series. Warsaw, 1935.

Author and Subject Index

Absorbing subset 4
Adjoint of a bounded linear
    operator 13
Alaoglu theorem 13
Algebra,
    semi-simple 7
    commutative 7
Arsove, M.G. 129
Arsove theorem 115
Arsove-Edwards theorem 123

B-algebra 7
    topology isomorphic 7
Ball, open 4
    unit 5
Banach, S. 54
Banach space 5
Banach theorem 13
Banach-Steinhaus theorem 9
Barrel 4
Barrel space (barrelled space) 4
Barrel theorem 8
Base, at a point 2
    local 3
Basis 3, 28, 117
    absolutely convergent 42
    block basis 65
    boundedly complete 36
    Cesaro basis 46
    dual generalized 118
    equivalent 62
    extended Markushevich 115
    first category basis 109
    generalized 115
    Hamel basis 3
    Markushevich basis 115
    monotone 35
    normal 79
    normalized 79
    orthogonal 111
    orthonormal 79
    retro-basis 34

Basis,
    Schauder basis 29, 117
    second category basis 109
    shrinking 34
    similar generalized 123
    T-Basis 45
    total generalized 115
    unconditional 38
    uniform 42
    weak 28
    weak* 28
    weakly uniform 34
Bessaga, C. 68, 78
Bessaga-Pelczynski theorem
    65
Bessel's inequality 6
Biorthogonal system 31, 114
    maximal 114
Boas, R.P 125
Body 23
Bounded subset 4
Bourbaki, N. 17

Cantor's triadic (-ternary) point set
    67
Cauchy net 5
Cauchy sequence 4
Characteristic function 1
Circled subset 3
Closed graph theorem 9
Closed set 2
Closure 2
Cluster point 2
Coefficient functionals of a basis
    29, 115
Coefficient mapping 115
Compact set 2
    conditionally compact set 2
    sequentially compact set 2
Conditionally weakly sequentially
    complete set 12
Conjugate space 10

Springer Tracts in Natural Philosophy